The Ethics Gap in the Engineering of the Future

The Ethics Gap in the Engineering of the Future: Moral Challenges for the Technology of Tomorrow

EDITED BY

SPYRIDON STELIOS
National Technical University of Athens, Greece

AND

KOSTAS THEOLOGOU
National Technical University of Athens, Greece

United Kingdom – North America – Japan – India – Malaysia – China

Emerald Publishing Limited
Emerald Publishing, Floor 5, Northspring, 21-23 Wellington Street, Leeds LS1 4DL

First edition 2025

Editorial matter and selection © 2025 Spyridon Stelios and Kostas Theologou.
Individual chapters © 2025 The authors.
Published under exclusive licence by Emerald Publishing Limited.

Reprints and permissions service
Contact: www.copyright.com

No part of this book may be reproduced, stored in a retrieval system, transmitted in any form or by any means electronic, mechanical, photocopying, recording or otherwise without either the prior written permission of the publisher or a licence permitting restricted copying issued in the UK by The Copyright Licensing Agency and in the USA by The Copyright Clearance Center. Any opinions expressed in the chapters are those of the authors. Whilst Emerald makes every effort to ensure the quality and accuracy of its content, Emerald makes no representation implied or otherwise, as to the chapters' suitability and application and disclaims any warranties, express or implied, to their use.

British Library Cataloguing in Publication Data
A catalogue record for this book is available from the British Library

ISBN: 978-1-83797-636-2 (Print)
ISBN: 978-1-83797-635-5 (Online)
ISBN: 978-1-83797-637-9 (Epub)

Printed and bound by CPI Group (UK) Ltd, Croydon, CR0 4YY

INVESTOR IN PEOPLE

'To our students, engineers and citizens of tomorrow'

Contents

About the Editors	*ix*
About the Contributors	*xi*
Acknowledgments	*xv*

Prolegomena: Mind the Gap! Now and for Ever   *1*
Spyridon Stelios and Kostas Theologou

Section 1: Artificial Intelligence

Chapter 1   Who Should Obey Asimov's Laws of Robotics? A Question of Responsibility   *9*
Maria Hedlund and Erik Persson

Chapter 2   Criminal Justice in the Age of AI: Addressing Bias in Predictive Algorithms Used by Courts   *27*
Rahulrajan Karthikeyan, Chieh Yi and Moses Boudourides

Chapter 3   Ethical Challenges in the New Era of Intelligent Manufacturing Systems   *51*
Emmanouil Stathatos, Panorios Benardos and George-Christopher Vosniakos

Chapter 4   From Croesus to Computers: Logic of Perverse Instantiation   *83*
Goran Rujević

viii   *Contents*

Chapter 5   The Gradual Unavoidable Colonization of the Lifeworld by Technology   *105*
Kostas Theologou and Spyridon Stelios

Section 2: Biotechnology

Chapter 6   Ethical Aspects of Promises and Perils of Synthetic Biology   *119*
Ivica Kelam

Chapter 7   Adapting (Bio)ethics to Technology and Vice Versa: When to Fight and When to Collaborate With Artificial Intelligence   *137*
Iva Rinčić and Amir Muzur

Section 3: Space

Chapter 8   Are Space Technologies Untimely?   *159*
Tony Milligan

Chapter 9   Moral Vistas to Xenic Beyonds: Fostering Moral Imagination to Pre-empt Monsterization in Future Encounters With Extraterrestrial Life   *177*
George Profitiliotis

Chapter 10   Planning for the Future in Space – With or Without Radical Biomedical Human Enhancement?   *201*
Rakhat Abylkasymova and Konrad Szocik

Chapter 11   Building Better (Space) Babies: Bioastronautics, Bioethics, and Off-World Ectogenesis   *215*
Evie Kendal

Index   *229*

About the Editors

Dr Spyridon Stelios is a Teaching and Research Associate at the National Technical University of Athens (Faculty of Applied Mathematical and Physical Sciences/Department of Humanities, Social Sciences and Law). He lectures at several schools of NTUA, and his main research interests are in ethics, epistemology, and experimental philosophy. He has authored several book chapters, articles in international peer-reviewed journals and conference proceeding in these areas. He also serves as a reviewer for several academic journals.

Kostas Theologou is a Professor at the Faculty of Applied Mathematical and Physical Sciences of the National Technical University of Athens (and Director of the Department of Humanities, Social Sciences and Law); he is also the Director of the journal Signum http://aked.ntua.gr/signum.html. He is the Vice Director of Studies of European Civilization studies and the Moderator of the module 'Social Theory and Modernity" of the School of Humanities of the Hellenic Open University. His research interests focus on the History and Philosophy of Culture (technology, urban space, collective identities, bioethics, biotechnology, etc.).

About the Contributors

Rakhat Abylkasymova, independent researcher, Poland. Her areas of interest are linguistics, international business, and future challenges affecting humanity, with a focus on our future in space. She has published in *the International Journal of Astrobiology* and *Space Policy*, among others.

Panorios Benardos currently serves as an Assistant Professor in the School of Mechanical Engineering of the National Technical University of Athens (NTUA) in the field of 'Intelligent Manufacturing". He received his Diploma in Mechanical Engineering in 2001 and his PhD in 2009 from NTUA. He has worked as a Lecturer and an Assistant Professor in the University of Nottingham from 2014 to 2018. He has participated in 21 nationally (GSRT, EPSRC, Innovate UK) and EU-funded research projects (FP7, H2020, HE). He has published 42 papers in international scientific journals and conferences (24 and 18 respectively) and coauthored 2 book chapters. His current research interests involve the application of artificial intelligence and machine learning methods in modeling and optimization of manufacturing processes and systems.

Moses Boudourides served as a Professor of Practice at the Arizona State University School of Public Affairs and is currently a member of the faculty of the Northwestern University School of Professional Studies Data Science Online Graduate Program. Previously, he held the position of Professor of Computational Mathematics at the University of Patras in Greece and the Department of Electrical and Computer Engineering at the Democritus University of Thrace, also in Greece. His expertise lies in applied and computational mathematics, network science, and computational social science.

Maria Hedlund is an Associate Professor at the Department of Political Science, Lund University, Sweden. Her research focuses on theories on democracy and the role of experts in democratic processes and in public administration, in particular regarding questions on emerging technologies, ethics, governance, responsibility, and legitimacy. Her recent publications include 'AI and epigenetic responsibility" (in *Epigenetics and Responsibility: Ethical Perspectives*, Anna Smajdor, Daniela Cutas, Emma Moormann & Kristien Hens (eds.), Bristol University Press, 2024), 'Ethicisation and reliance on ethics expertise" (*Res Publica*, April 2023), with Erik Persson 'Expert responsibility in AI development" (*AI & Society*, June 2022), and 'När maskiner fattar beslut – vem är ansvarig?" (When machines make decisions – who is responsible?) (in *Statsvetenskaplig tidskrift* 122, 2020).

xii   About the Contributors

Rahulrajan Karthikeyan is a graduate student in the Computer Systems Engineering (CSE) program at the School of Computing and Augmented Intelligence, Arizona State University. He has worked as a Machine Learning Research Assistant in the Cooperative Robotics Systems Lab (CRS), collaborating with PhD students on research involving multi-modal robot communication, temporal-spatial inverse semantics, and various projects leveraging Inverse Reinforcement Learning, ROS, and Computer Vision. In addition, he co-founded the Google Developer Student Club (DSC) at ASU, where he has taught Flutter, TensorFlow, Machine Learning, and Computer Vision to over 200 Bachelor's and Master's students at each DSC event. His accomplishments at ASU include winning more than 20 hackathons and coding challenges. He has also collaborated with Professor Boudourides on several machine learning projects.

Ivica Kelam, PhD, Head of the Department of Philosophy and History at the Faculty of Education, teaches the philosophy of education, ethics of the teaching profession and bioethics. Since 2018, he has been the Head of the Osijek Unit of International Bioethics Chair. In 2014, he successfully defended his doctoral thesis entitled Genetically modified crops as a bioethical problem. Since 2016, he has been the Head of the Center for Integrative Bioethics at J.J. Strossmayer University of Osijek. Since 2020, he has taught the bioethics of sports and sociology of sports at the Faculty of Kinesiology, teaches ethics in biosciences at the doctoral study in molecular biology at the University of Osijek, and teaches bioethics at the doctoral study at the Faculty of Education. He is the President of the Organization Committee of the Osijek Days of Bioethics, an international bioethical conference held in Osijek. He has published the book Genetically Modified Crops as a Bioethical Issue, over 40 scientific papers in Croatian and foreign scientific journals, and participated in more than 80 scientific conferences in Croatia and abroad. In 2020, he was elected President of the Croatian Bioethical Society.

Evie Kendal is a Bioethicist and Public Health Scientist whose work focuses on emerging reproductive technologies. She is the Head of the Ethical, Legal, and Social Implications of Emerging Technologies (ELSIET) research group of the Iverson Health Innovation Research Institute at Swinburne University of Technology and a Senior Lecturer of health promotion in the School of Health Sciences. She is the author of 'Ectogenesis and the Ethics of New Reproductive Technologies for Space Exploration" (in *Transhumanism*, Springer 2022), 'Biological Modification as Prophylaxis: How Extreme Environments Challenge the Treatment/Enhancement Divide" (in *Human Enhancements for Space Exploration*, Springer 2020), and 'Desire, Duty and Discrimination: Is There an Ethical Way to Select Humans for Noah's Ark?" (in *Reclaiming Space*, Oxford University Press 2023).

Tony Milligan is a Research Fellow in Ethics with the Department of Theology and Religious Studies at King's College London. His publications include *The Ethics of Political Dissent* (2023), *Nobody Owns the Moon* (2015), and the coedited volume *The Ethics of Space Exploration* (2016). He is currently working on a short introduction to societal issues concerning space: *What is Space Exploration For?* with Koji Tachibana, and a SpringerBrief on *Indigenous Cosmological Visions of the Pleiades*.

About the Contributors **xiii**

Amir Muzur is a Full Professor (since 2013) and the Head of Department of Social Sciences and Medical Humanities (since 2008) at University of Rijeka – Faculty of Medicine. His major fields of research/teaching interest are history of medicine and bioethics and theoretical neuroscience. He authored/coauthored more than 30 books (local history, history of medicine, popularization of neuroscience, essays, travelogues, poetry) and about 700 articles in scientific and other periodicals and participated in more than 200 conferences (more than 40 invited lectures). Amir Muzur was the President of Croatian Bioethics Society (2012–2016) and now is the Director of the Fritz Jahr Documentary and Research Centre for European Bioethics at University of Rijeka (since 2013).

Erik Persson received his PhD in Philosophy from Lund University in 2009. He is currently an Associate Professor in Practical Philosophy at the Department of Philosophy, Lund University and Researcher at RISE – Research Institutes of Sweden. His previous employers include Umeå University, the Nordic Genetic Resource Centre and the Institute for Theological Inquiry in Princeton. His research and teaching are primarily focused on applied ethics, value theory, environmental ethics, space humanities, and the philosophy of AI and other emerging technologies. His recent publications include 'The future of AI in our hands? To what extent are we as individuals morally responsible for guiding the development of AI in a desirable direction?" (*AI & Ethics* 2: 683 – 695), co-authored with Maria Hedlund.

George Profitiliotis is a Research Scientist affiliate at Blue Marble Space Institute of Science and a speculative fiction author serving the New Weird. He is an electrical and computer engineer, further trained in interdisciplinary environmental studies, biomimetics, and strategic innovation management, and holds a PhD on the application of environmental economics to planetary protection policy. He has worked extensively in the fields of strategic foresight, futures literacy, and futures studies. During his first PostDoc, he studied the pertinence of anticipation to the search for extraterrestrial life from the viewpoints of particular subfields of the humanities, social sciences, and policy, with a focus on the proactive management of a future discovery. His second PostDoc focused on anticipating the future ethical, legal, and societal implications of quantum technologies.

Iva Rinčić graduated in 2000 in Sociology and Croatian Culture from the Department of Croatian Studies at the University of Zagreb and defended her Master's thesis in 2005 at the Faculty of Political Sciences. She defended her PhD thesis 'Theoretical Strongholds, Achievements and Perspectives of Bioethical Institutionalisation in the European Union' in 2010 at the Faculty of Philosophy in Zagreb. Since 2001, she has been working at the Department of Social Sciences and Medical Humanities of the Rijeka University Faculty of Medicine (since 2022 as Full Professor), and since 2014, also at the Department of Public Health of the University of Rijeka Faculty of Health Studies. She has participated at more than 60 Croatian and international conferences, published approximately 70 scientific and professional articles, and five books. From 2013 till 2021, she was the Director of the University of Rijeka Foundation.

xiv   *About the Contributors*

Goran Rujević is an Assistant Professor at the Department of Philosophy of the Faculty of Philosophy in Novi Sad. He earned his Bachelor's and Master's degrees in Philosophy from the Faculty of Philosophy in Novi Sad, where he also defended his PhD thesis on Kant's philosophy of science in 2019. His main areas of interest are philosophy of natural sciences and argumentation theory, but he also has interest in issues regarding teaching and popularizing philosophy. Apart from scholarly work, he often translates works from English and German into Serbian; of note is his work on translating Nick Bostrom's seminal book *Superintelligence* into Serbian.

Emmanouil Stathatos is a Mechanical Engineer who graduated from the National Technical University of Athens (NTUA) in 2006. He earned his PhD in 2020, focusing on the optimization of the Selective Laser Melting (SLM) process through numerical modeling and machine learning techniques. Since 2009, he has served as a Technical Consultant to Greek industrial firms. Currently, he is a Postdoctoral Researcher at the Manufacturing Technology Lab, NTUA, where he contributes to Horizon Europe and Erasmus+ projects. His teaching roles have included Laboratory Teaching Assistant for Computer Aided Manufacturing, Additive Manufacturing, and Discrete Event Simulation for manufacturing systems, as well as Instructor for the Advanced Manufacturing Systems course in the Interdepartmental Master's Program on Automation Systems. He has published 11 papers in international scientific journals and conferences. His research is dedicated to integrating AI into manufacturing to redefine the paradigms of smart production.

Konrad Szocik, University of Information Technology and Management in Rzeszow, Poland. From 2021 to 2022, he was a Visiting Fellow at the Yale Interdisciplinary Center for Bioethics. His interests include ethics, philosophy and bioethics of space exploration, feminism, antinatalism, procreative ethics, and population ethics. He is the author of *The Bioethics of Space Exploration* (Oxford University Press, 2023) and *Feminist Bioethics in Space* (Oxford University Press, 2024).

George-Christopher Vosniakos studied Mechanical Engineering at the National Technical University of Athens and obtained an MSc in Advanced Manufacturing Technology in 1987 and a PhD in Intelligent CAD-CAM interfaces in 1991 from UMIST-Britain. He has worked in the software industry in Germany and as a Lecturer at UMIST. In 1999 he joined NTUA, where he is currently Professor of Manufacturing Systems and director of the Manufacturing Technology Laboratory. He has authored 110 Journal and over 108 Conference papers obtaining 2,850 Scopus citations. He is editorial board member of six major International Journals. His recent research interests include AI in Manufacturing.

Chieh Yi is a graduate student in the Computer Systems Engineering (CSE) program at the School of Computing and Augmented Intelligence at Arizona State University. He has collaborated with Professor Boudourides on several machine learning projects and is actively involved in the ASU EPICS program, where he contributes to innovative technical projects.

Acknowledgments

Special thanks to Katy Mathers, Lauren Kammerdiener, and Brindha Thirunavukkarasu from Emerald for their help, encouragement, and support as well as to the anonymous reviewers for their insightful comments.

Prolegomena: Mind the Gap! Now and for Ever

Spyridon Stelios and Kostas Theologou

National Technical University of Athens, Greece

This book discusses Technology, Engineering, and Ethics and the ways they may be related to two strictly mutually defined theoretical concepts: the "gap" and the "future". Both concepts have not occupied, as much as we think they should, the research around Engineering ethics and the Philosophy of technology. Nevertheless, they are particularly important as they express both the dizzying technological progress (see "future") and the side effects of this forward march (see "gaps").

What is the future? One could respond that it is something that (in any case) does not exist as it cannot be experienced as such – it is experienced only as present. It is a mental, immaterial, functional aspiration that we use to give meaning to our existence, to make choices and plan our actions. As a mental construct or merely a perception, the future is an extremely fruitful and therefore useful field of thought guiding our behavior. But does it have any value beyond its functionality? Or rather, what can we think of a concept which is also only a thought, which is on the same sphere as dreams and imagination and which has no tangible empirical manifestation? The answer is "a lot". The future offers fertile ground for philosophical inquiry, the preeminent mental pursuit that humans developed and which led to the birth of science itself. And it is this philosophical examination of a nonempirically existent field of reference that can transform or change the present.

The swift leaps of technology create inductively documented images of the future of humanity. Technology is threaded together with the concept of progress leading to a better future, so the ethics of technology is interwoven, above all, with future representations. We are curiously interested in what happens, e.g., with generative intelligence today, but we are more interested in what will happen tomorrow regarding that and similar technologies. The future is omnipresent in the present/here, or more precisely, the present/here is constantly in the future. This collective obsession undoubtedly awakens people to potential dangers, but at the same time prepares them to accept innovations and interventions more easily in their lifeworld. Let us not forget also that there is always the risk any discussion

---

The Ethics Gap in the Engineering of the Future, 1–5
Copyright © 2025 Spyridon Stelios and Kostas Theologou
Published under exclusive licence by Emerald Publishing Limited
doi:10.1108/978-1-83797-635-520241001

2 *Spyridon Stelios and Kostas Theologou*

of the future makes it easier to avoid discussion and criticism of the state of the present.

Regarding technology, ethical reflection is limited by a multiplicity that derives from the object of study itself and which creates a kind of gap. But why refer to a gap? What is missing from our moral concerns, intuitive and explicit, that needs to be filled in? What might largely led to the conception and creation of this book's theme in the first place? To begin with, the moral gap is somehow fed by and related to the future dimensions of technology. No one can predict exactly what specific ethical issues will arise from tomorrow's amazing innovations and the ongoing digital transformation of our civilization. As much as we have an image of the future through the technological present, this image and sense has already contaminated, altered, and changed the course of development for every artefact. In fact, it could be argued that any assessment adds an extra boost to the final form of the "new" to come; and the new always shocks upon appearance the bystanders, either in the form of art or artefact. Simply put, what we expect, because we expect it, will be even further ahead, even more sophisticated. This does not involve a specific timeframe. When we refer to the appearance/evolution of an artefact in the future, we are not referring to temporal estimates about when it will come. It is not worth much, in philosophical terms, to say that, e.g., in 2050 there will be vehicles that can drive on the road, dive in water or fly (as, for example, in the 1960s the author Philip K. Dick predicted in his novel "Do Androids Dream of Electric Sheep?" that in 1992 there would be hovercars). The assessment of the future of any technological product associates with the product's final form, regardless of when this format will be completed. It is the form that we set as final and beyond which we cannot think of another.

Of course, one would reasonably argue, this is completely subjective. Another form I believe cars will have—keeping the essence of what it means to be a car—and another someone else. In fact, the possibilities, and forms, are endless. And not only. Their very future perspective, expressed through these forms, geometrically evolves their possible final state. In other words, the more we think about the *end* (*telos*) of a technological phenomenon, the further it draws away! This explosive expansion of future projections of the technological phenomenon creates gaps in their ethical processing. The nature of technology creates a kind of never reached imaginative infinity within our everyday lives. Against this, moral philosophers try, but rather fail, to find suitable clothes for artificial children who never stop growing.

Today's technology also seems to have gaps, especially if it is seen in its holistic social dimension. Technology is not a uniform phenomenon that satisfies the needs and desires of every culture to the same extent. Also, if we carefully examine today's innovations, we have the impression that they have not spread to all areas of our lives equally. At least not yet. For example, developments in the field of biotechnology seems fewer and to a lesser extent than those in the field of artificial and computational intelligence. Space technology is also developing less than artificial intelligence and biotechnology. Of course, the reasons for these development gaps are many (see for instance, economic) and certainly cannot be analyzed in this short introduction. Perhaps it is naive even to talk about categorization and a common ground between different artefacts that allows their comparison. What could be argued is that the larger the network of people and organizations involved in a field, the greater will be its rate of development. In the case of space exploration technology, for instance, the players are not really that many. There are a few dozen countries in the world that have and are developing this kind of technology.

In this book, the concept of the future is examined in three general technological fields to highlight possible gaps in their moral orientation. These three thematic sections are: artificial intelligence, biotechnology and space. Of course, the multifactorial dimension of the topics under discussion inevitably leads to overlaps between the sections.

Section 1: Artificial Intelligence provides general reflections on the ethical nature of AI. This part is composed of five chapters. The first chapter, entitled "Who should obey Asimov's laws of robotics? A question of responsibility" by Maria Hedlund and Erik Persson, discusses the safety value of implementing Asimov's Laws of Robotics as a future general framework that humans should obey. Within this framework, the implementation of the law in human legislation is being considered, especially regarding people or companies that develop, build, market, or use AI, now and in the future. The second chapter, entitled "Criminal justice in the age of AI: Addressing bias in predictive algorithms used by courts" by Rahulrajan Karthikeyan, Chieh Yi, and Moses Boudourides focuses on recent studies aimed at mitigating biases in algorithmic decision-making within the realm of predictive policing algorithms employed in the criminal justice system. The authors reassess recidivism rates and implement adversarial debiasing in conjunction with fairness metrics. In "Ethical challenges in the new era of intelligent manufacturing systems", our NTUA colleagues Emmanouil Stathatos, Panorios Benardos, and George-Christopher Vosniakos explore the ethical challenges arising from the integration of advanced AI technologies into intelligent manufacturing systems. The chapter reflects on issues such as data privacy, job displacement, the impact of automation on workforce dynamics, and the psychological effects of working alongside AI-powered systems. The fourth chapter, entitled "From Croesus to computers: Logic of perverse instantiation" by Goran Rujević, analyses old and new tales of perverse instantiation arguing that is fundamentally a philosophical in nature, problem so old that even Socrates had to face it. The author aims at uncovering this fundamental problem and show its connection to the contemporary field of AI ethics, especially when such failure mode presents an existential risk (see superintelligent AI). In the last chapter of the section, entitled "The gradual unavoidable colonization of the lifeworld by technology" by Kostas Theologou and Spyridon Stelios, the technological colonization of the lifeworld is being discussed, introducing the concept of tech-lifeworld.

Section 2: Biotechnology focuses on specific ethical issues raised by human activity in the field of biotechnology. This part consists of two chapters. The first chapter, entitled "Ethical aspects of promises and perils of synthetic biology" by Ivica Kelam, investigates the phenomenon of synthetic biology through an ethical analysis of the unfulfilled promises and potential perils surrounding this

4 Spyridon Stelios and Kostas Theologou

technology. At first, he discusses the problem of defining the inter-disciplinary field of synthetic biology and then proceeds with a brief history of systemic biology and the groundbreaking creation of Synthia, the first synthetic organism. The potential benefits of synthetic biology are then discussed together with ethical and regulatory issues. In "Adapting (bio)ethics to technology and vice versa: when to fight and when to collaborate with artificial intelligence", Iva Rinčić and Amir Muzur investigate the possibility of a systematic study of adaptations human society will have to consider guaranteeing the obeyance to the fundamental ethical values within an environment of rapid advancement of AI. This possibility emerges as particularly important if we consider recent developments such as the creation of computer-generated synthetic lifeforms displaying characteristics of living beings. Against these contemporary developments, the authors propose a new discipline, *Epharmology*, as a methodological approach that could provide a systematic ethical framework.

Section 3: Space explores ethical challenges that might emerge through our greater familiarity with extraterrestrial reality. Space offers fertile ground for the discussion of existing and new ethical conceptualizations. This part consists of four chapters. The first chapter entitled "Are space technologies untimely?" by Tony Milligan provides an overview of fears and suspicions concerning the development and expansion of space technologies, classifying them as instances of "space skepticism." It goes on to argue that the technologies in question are not untimely with respect to the major challenges facing humanity. The second chapter entitled "Moral vistas to xenic beyonds: Fostering moral imagination to pre-empt monsterization in future encounters with extraterrestrial life" by George Profitiliotis offers a theoretical exploration and a practical intervention, in line with pragmatist ethics, in the form of a novel futures literacy workshop to help pre-emptively decrease the potential for the monsterization of humans and extraterrestrial life in the case of a future discovery. The workshop is envisioned as a preparatory, complementary pedagogical approach to the traditional teaching of applied ethics to university students. The third chapter, entitled "Planning for the future in space—with or without radical biomedical human enhancement?" by Rakhat Abylkasymova and Konrad Szocik, discusses the possibility of long-term space exploration requiring extraordinary solutions, such as the possible obligation or requirement to apply radical human enhancement. Furthermore, they refer to the feminist perspective and to issues such as exclusion and power structures. The final chapter, entitled "Building better (space) babies: Bioastronautics, Bioethics and off-world ectogenesis" by Evie Kendal explores some of the ethical issues surrounding ectogenesis—the development and use of artificial womb technology—and its space applications in establishing an off-world human society.

Advances in several technological sectors have made engineering a rapidly and dynamically expanding industry. Every invention concerning robots, Bioengineering, Space, and other fields is now being developed, perfected, and implemented in a fast-track pace. Generally, there is not enough time to experiment, to test, or to approve technical innovations. This is how markets push consumers to novel risks creating gaps in their effort and process of integrating these fascinating and not always necessary artifacts. This dynamic also gives insight on technology's future impact both on the individual and on the collective/social level. We are witnessing technology transforming itself into new potential and sometimes hypnotic applications. We do not only see it transform but we feel rather unprepared to adjust our existence, our being, within its ever-altering framework. The future is being shaped up in an imperceptible manner now and requires proper moral treatment and provisions. This process is better comprehended by looking at future scenarios and challenges. For example, imagine in the future waking up in a colony on the planet Mars and having to decide how to manage an impending biological threat using AI-based systems and biotechnological knowledge. This kind of approach is generally taken in the chapters.

This book infuses applied ethics and engineering in three technological domains to ferment and stimulate ethical reflections on their future development. The reflections and insights that the 11 chapters provide are key to understanding the change our personal, everyday life is undergoing. In particular, the fifth chapter seeks to collect all the threads spun by the different contributions to enlighten the thesis of a technological colonization of the lifeworld. This technological colonization of lifeworld serves as a link between the chapters and themes.

So, the technological development is moving so far that it is difficult to timely and properly adjust culturally-socially and personally. This fast-moving pace leads to a diversity of ethical aspects of the future visions of life and artificial–biological interactions. Based on this insightful and incisive look into tomorrow, engineers, philosophers, intellectuals and scientists as well as researchers, educators, and scholars in the fields of technology and humanities will likely benefit from reading the chapters of this collective volume finding theoretical and practical guidelines.

Section 1

Artificial Intelligence

Chapter 1

Who Should Obey Asimov's Laws of Robotics? A Question of Responsibility

Maria Hedlund and Erik Persson

Lund University, Sweden

Abstract

The aim of this chapter is to explore the safety value of implementing Asimov's Laws of Robotics as a future general framework that humans should obey. Asimov formulated laws to make explicit the safeguards of the robots in his stories: (1) A robot may not injure or harm a human being or, through inaction, allow a human being to come to harm; (2) A robot must obey the orders given to it by human beings except where such orders would conflict with the First Law; (3) A robot must protect its own existence as long as such protection does not conflict with the First or Second Law. In Asimov's stories, it is always assumed that the laws are built into the robots to govern the behaviour of the robots. As his stories clearly demonstrate, the Laws can be ambiguous. Moreover, the laws are not very specific. General rules as a guide for robot behaviour may not be a very good method to achieve robot safety – if we expect the robots to follow them. But would it work for humans? In this chapter, we ask whether it would make as much, or more, sense to implement the laws in human legislation with the purpose of governing the behaviour of people or companies that develop, build, market or use AI, embodied in robots or in the form of software, now and in the future.

Keywords: The laws of robotics; Asimov's laws; robot ethics; AI ethics; safety; responsibility; democracy

Introduction

The aim of this chapter is to explore the value of implementing Asimov's Laws of Robotics as a general framework for humans with the purpose of governing the behaviour of people or companies that develop, build, market or use artificial

---

The Ethics Gap in the Engineering of the Future, 9–25

Copyright © 2025 Maria Hedlund and Erik Persson

Published under exclusive licence by Emerald Publishing Limited

doi:10.1108/978-1-83797-635-520241002

intelligence (AI), embodied in robots or in the form of software, now, and in the future. In 1942, the science fiction author Isaac Asimov introduced the Laws of Robotics to make explicit the safeguards of the robots in his stories (Asimov, 1942). Generally, safety refers to the prevention of harm or other non-desirable outcomes (Hansson, 2012). Asimov's laws aimed to do exactly that:

(1) A robot may not injure a human being or, through inaction, allow a human being to come to harm.
(2) A robot must obey orders given to it by humans except where such orders would conflict with the First Law.
(3) A robot must protect its own existence as long as such protection does not conflict with the First or Second Law (Asimov, 1942).

Later, Asimov added a fourth, or 'Zeroth' Law, that preceded the other laws in terms of priority:

(0) A robot may not injure humanity or, through inaction, allow humanity to come to harm (Asimov, 1985).

The laws are intuitively appealing: they are simple and straightforward, and they embrace essential ethical principles of many societies. However, as his stories clearly demonstrate, the Laws can be ambiguous, making it difficult for the robot to follow them. Moreover, the laws are not very specific. For instance, should a robot obey orders from any human? Or how would a robot act if information is kept from it? In addition, these laws were hardwired into the robots' 'positronic' brains. It is doubtful whether it would be possible to build general laws like these into a real robot. And if it would, who should be responsible for the actions of the robot?

Responsibility presumes the capability to make a difference and awareness of what you are doing. Currently, robots lack these attributes and cannot be responsible for their actions, which is a good reason not to hand over complex decisions to robots. Adding context specific considerations to general rules, such as the Laws, are necessary to make reasonable judgements in real-world situations (Persson & Hedlund, 2024). Although self-learning machines get increasingly better at reading the environment in which they act, they do not understand what they are doing, why they are doing it, or the consequences of what they are doing. Thus, general rules for robot behaviour may not be a very good method to achieve robot safety – if we expect robots to follow them. But would they work for humans?

We are not the first to ask this question. In light of the rapid AI development that we are currently witnessing, there is concern for the human relationship to more and more capable robots. And Asimov's laws seem to have encouraged political actors as well as roboticists and software developers: a search on 'three laws of robotics' in Google scholar in December 2023 gives more than 607,000 hits.

Two examples of how political actors have been inspired by Asimov's laws are South Korea and the European Union. In 2007, the South Korean Government drew up a Robot Ethics Charter that would cover ethical standards to be programmed into robots. According to an official from the Ministry's Code of Ethics, the charter would reflect Asimov's Laws of Robotics. However, in a revised version of the South Korean Robot Ethics Charter from 2018, Asimov's laws are not mentioned (Choi et al., 2019).

In 2016, the European Parliament suggested acknowledging robots as 'electronic persons' and that Asimov's Laws of Robotics should be guiding ethical principles for robots (EP, 2016a). The suggestion met hard criticism for the misinterpretation that fictional laws intended for robots could protect humanity (EP, 2016b, 2017), and did not recur later in the process of developing ethical guidelines for AI. These examples from the political sphere illustrate that Asimov's laws have an impact outside the fictional world.

Also in the scholarly literature, we find the idea to apply Asimov's Laws. One example is Clarke (1993, 1994), who sees Asimov's Laws as a set of principal guidelines, or lessons, to be applied during design, development and use of robotic systems. One such lesson is that Asimov's Laws do not designate any particular class of humans as more deserving, that is, they should benefit all humans equally. Another lesson is that we must not focus only on the technology as such but also on how the technology is used.

Another example of the impact of Asimov's Laws on scholarly literature is Murphy and Woods (2009), who suggest three alternative laws. Their alternative laws place the responsibility for robot safety on humans. They also suggest that the hierarchy with humans as superior and robots as subordinate is not always suitable. For instance, we might prefer that the robot ignores a hacker.

In a more recent article that takes Asimov's Laws as an explicit point of departure, Balkin (2017) argues that what we need is laws directed at the people who programme and use robots, AI agents and algorithms. Balkin's proposal focuses on trust and fairness and states that AI businesses should have an obligation to be trustworthy towards their end users and to the public, and that algorithm operators have a duty not to externalise the costs of algorithmic decision-making onto others.

The connection of this proposal to Asimov's Laws is emphasised by Pasquale (2017), who suggests a Zeroth Law to 'ensure the viability of Balkin's three laws'. With reference to Microsoft's chatbot Tay, which quickly adapted its messages in a racist and misogynist direction, Pasquale suggests that the creator of a robot should be obliged to build in constraints on the code's evolution. This has some similarity to Asimov's Laws, which were hardwired into the robots.

Another kind of reference to Asimov's Laws in the scholarly literature is how they are included or excluded from reviews of ethical guidelines on AI. In the context of reviewing current frameworks for regulating AI, all published between 2016 and 2019, Torres and Penman (2021) include Asimov's Laws from the 1940s as one of these frameworks of AI. They found that Asimov's Laws 'easily [. . .] matched the frameworks of today's AI mainstream'.

12 Maria Hedlund and Erik Persson

In another review, Hagendorff (2020) explicitly excludes Asimov's Laws from the ethical guidelines under scrutiny. This review only includes guidelines published within the last five years, and the fact that the author justifies why he excludes Asimov's fictional laws from the 1940s is a further sign of their influence in thinking about AI development.

There are also examples of practices that claim to consider Asimov's Laws. According to Abdullah et al. (2021), medical bioethical research 'has always considered the Asimov laws, no matter how primitive they were, in the bioethical design of medical AI', and Kaminka et al. (2017) use Asimov's laws to examine safety and autonomy in molecular robots fabricated from a technique called DNA origami.

It is clear that we are not the first to ask whether Asimov's Laws could work for humans. However, we believe that it is necessary to try them out in some real-case situations. We also believe that it is necessary to incorporate the concept of 'responsibility' to do so. The remainder of this chapter is structured as follows. First, we will outline an understanding of responsibility that put emphasis on control and awareness. After that, we present our proposed idea of Asimov's Laws directed at humans, and apply it on a hypothesised real-case contemporary situation. This analysis helps us to finetune the Laws directed at humans. Next, we apply the revised version of the proposition on the Laws directed at humans on three significantly different and morally relevant directions of AI development. Finally, we present our conclusions and discuss potential ways forward.

Responsibility

In this chapter, we will focus on two aspects of responsibility that has particular relevance for our thought experiment: awareness and control. Normally, we are morally responsible for something we have caused as long as we are not acting under coercion or ignorance and are aware of the moral nature of the action (or inaction), that is, that we have moral agency (Held, 1970; Sneddon, 2005; Thompson, 1987).

Adequate knowledge about the causal relations and consequences of an action is necessary for any understanding of responsibility (Adam & Groves, 2011; Thompson, 1987). Of relevance is to what extent the agent is able to realise the effects of an action, the effects of not to act, or the effects of acting differently (Fischer & Ravizza, 1998). This is also valid for knowledge about right and wrong. In some situations, what is right and wrong is not contestable, but in many cases, there are no definite rights or wrongs. However, even without any definite answer on the question of right and wrong, there are always context dependent norms of what constitute a right or a good behaviour (Hedlund, 2012). Bad actions resulting from (real or alleged) ignorance of moral norms are blameworthy and can be seen as paradigm cases of moral responsibility (FitzPatrick, 2008). An actor who unintentionally or unvoluntary has caused a bad situation can be causally but not morally responsible, as she has not done anything blameworthy or objectionable (Talbert, 2008; Thompson, 1987).

To be responsible, agents must be able to control their actions (or non-actions) (Fischer, 1982). 'Control' is often referred to as the power to determine whether or not something occurs (Kane, 2002). In relation to AI algorithms that may develop in directions that is difficult or impossible to predict, even for the designer of the algorithm, the concept of 'responsibility gap' has been suggested to denote situations in which human control is undermined (Matthias, 2004). A responsibility gap is however an unwelcome situation, both practically and conceptually. We do not want to find ourselves in a situation in which AI gives rise to harm without being able to make someone accept that they are responsible for this harm. Perhaps it is too much to demand that the designer should be able to control every causal step in what the AI does. Conceptually, 'responsibility' does not seem to do its job here. Perhaps 'control' is too strong a requirement for responsibility, at least in cases involving AI algorithms and other autonomous systems.

To avoid a responsibility gap, Himmelreich and Köhler (2022) suggest that we instead build responsibility not on control, but on another kind of causal-like relation such as supervision. Drawing on Nyholm (2018), who argues that the relation between the developer and the AI is analogous to the relationship of supervision that is ongoing between a parent and a small child, Himmelreich and Köhler (2022) contend that developers stand in a supervisory relation to AI and autonomous systems. They have control in the sense that they 'maintain, improve, and teach the AI system what to do and how to behave' (Himmelreich & Köhler, 2022). With this weaker relationship between the developer and the AI, the developer can be responsible for what the AI does, even though she cannot fully predict or control its exact course of action (Nyholm, 2018). Supervision places the incentive correctly, as the developer is the one who has influence over the AI by training it, and by this weaker causal-like relation, the developer can be responsible for a harm that an AI causes because she trained it (Himmelreich & Köhler, 2022).

Asimov's Laws of Robotics Directed at Humans

We suggest that a reasonable way to make Asimov's Laws apply to humans is to phrase them in terms of responsibility. Given these premises, Asimov's Law's directed at humans could look like this:

(1) AI developers have a responsibility to see to it that an algorithm may not injure a human being, or, through inaction, allow a human being to come to harm.
(2) AI developers have a responsibility to see to it that an algorithm obeys the orders given to it by humans except where such orders would conflict with the First Law.
(3) AI developers have a responsibility to see to it that an algorithm protects its own existence as long as such protection does not conflict with the First or Second Law.

14   *Maria Hedlund and Erik Persson*

And the Zeroth law:

(0) AI developers have a responsibility to see to it that an algorithm may not injure humanity or, through inaction, allow humanity to come to harm.

Could the laws, formulated like this, help remedy the damage that recommender algorithms have on the functioning of democracy and thereby avoid severely harming humans? Asimov's Laws aimed at preventing harms to humans and humanity. As pointed out by several scholars (e.g., Clarke, 1993; Schurr et al., 2007), 'harm' is notoriously vague and need to be specified. Schurr et al. (2007) chooses to operationalise 'harm' as 'a "significant" negative loss in utility', capturing as well physical as mental harm. Harm can be direct or indirect. One way of severely harming humans is to damage the democratic system. To try out our idea of applying Asimov's Laws on humans, we consider one particular harm, namely how recommender algorithms may damage the functioning of democracy.

Democracy here refers to a system in which citizens have equal rights and real possibilities to decide on common matters. In its ideal form, democracy is a clever way to reach agreements or make compromises when we do not agree on matters. Instead of fighting, we vote, and we discuss and deliberate. Even though a sound democracy requires that we accept that people have different views, some common ground is necessary for meaningful democratic debate. However, recommender algorithms contribute to giving us very different world views. Based on what we 'like' or share on social media platforms, or on what we search for in search engines, recommender algorithms direct us to content that we are assumed to prefer. The effect is that we are confronted with different world views, which make democratic debate and mutual tolerance difficult (Hedlund & Persson, 2024). Could Asimov's Laws directed at humans help remedy this damage to democracy and thereby protect humans from harm?

Like we stated above, our discussion refers to AI, embodied in robots or in the form of software (algorithms). Later, in the imagined future scenarios, we will primarily talk about super-intelligent AI. In this first case, we will only mention algorithms. We assume that relevant humans are those who design, develop and provide the algorithms. Humans can have several tasks in relation to an AI system, such as designers, developers, operators, deployers or users. The point here is that we are talking about humans involved in development of robots and systems built on AI technology, irrespective of the precise role they play. For the sake of brevity, we use 'developer' to denote any relevant human involved in development of AI, and 'user' for relevant humans that somehow make use of the technology.

Application on a Hypothesised Real-Case Contemporary Situation

Consider a developer of a recommender algorithm. What could these laws imply for her? We need to keep in mind that there is a hierarchy between the laws, with the Zeroth law as the superior law. Hence, first and foremost, the developer has the responsibility not to harm humanity.

We assume that damaging the function of democracy is a harm to humanity. We also assume that it is important for all humans, for different reasons, including a well-functioning democracy, to be exposed to other opinions and not only be exposed to the same opinions that you already have, which recommender algorithms tend to do. Thus, according to the Zeroth Law, the responsibility of the developer is to see to it that the algorithm does not cause different world views for different people, which would damage the function of democracy and thereby harm humanity. That would require that the algorithm does not give dissimilar recommendations to different people who are making the same search.

How would this align with the First, Second and the Third Law?

The First Law states that a human being may not be harmed. Hence, the developer has the responsibility to see to it that individual humans are not harmed. While individual human beings may benefit from a functioning democracy, they may suffer from getting search results that are not individualised. This is a harm that the developer (possibly) is responsible to avoid.

In 'Liar', Asimov (1941) discusses a similar but less dramatic case when a telepathic robot lies to people, since it has realised that the truth would be painful for them, but, in fact, the robot causes more harm for the people by lying to them. This parallels our algorithm case. By unreflectingly providing those who use the algorithm with what they want, they are probably instantly happy, but in the long run, they will get harmed by not taking part of ideas that differs from their own. Recall J. S. Mill's idea of the importance to expose one's opinions to critical scrutiny both for the good of the society that otherwise would stagnate in old habits and for individuals to thrive (Mill, 1859/2011).

But the First Law is subordinate to the Zeroth Law, aimed at protecting humanity from harm, and in this case, the harm to the individual human being cannot be avoided if humanity should be protected (again, assuming that damage to democracy is a harm to humanity). Harming the individual may however have effects that is detrimental for democracy in other ways. For instance, this individual may lose her confidence in the digital infrastructure that aims to serve a favourable discussion atmosphere and thereby the functioning of democracy.

Confidence, or trust, is a highly prioritised value in discussions on human-technology interaction and include aspects such as safety and transparency of AI systems (AI HLEG, 2019; Buruk et al., 2020; Fjeld et al., 2020), but as Duenser and Douglas (2023) argue, it is important to acknowledge that trust in AI involves not only reliance on the system itself, 'but also trust in the developers of the AI system'. In Balkin's (2017) words, AI developers 'have duties of good faith and trust towards their end-users'.

In our case, the individual's loss of trust in the AI system may, potentially, be outweighed by the individual's trust in the developer, given that it is known to the individual that the developer prioritises the superior law with the aim at protecting humanity. This is however not a very strong claim, as individuals tend to be self-interested and short-sighted. On the other hand, would it not be for the protection of humanity (that is, democracy), there would be no public debate at all, which would give the individual at least a good reason to trust the developer. Thus, as the hierarchies between the Laws imply, the developer has the responsibility to prioritise the long-term and collective over the short-term and individual.

How about the Second Law? According to this law, the responsibility of the developer is to see to it that the algorithm obeys human orders. While the law does not specify which human the algorithm should obey, we do not know whether it should obey the developer or the user.

In some sense, the algorithm has no choice but to follow the instructions of the developer. On the other hand, machine-learning algorithms may develop in a way that the developer cannot predict. Depending on user input, the algorithm adapts. This adaptation could perhaps be seen as a kind of obedience of the user. But if this adaptation of search results is the root to the damage to democracy, then the developer, according to the Zeroth Law, is responsible not to design adaptive algorithms, *or* to design adaptive algorithms that develop in another direction than creating diverse world views for different users. In the former case, we have ruled out the user as a human that the algorithm should obey. In the latter case, the user is still a human that the algorithm should obey. However, this means we will have to consider also users that are malevolent and for some reason want to utilise the algorithm to make harm or to destroy the algorithm.

We do not want the algorithm to obey all users equally. To avoid that, we suggest that we discriminate what we could call authorised users, that is, users who apply the algorithm as intended, and unauthorised users, such as hackers. But does not the premise in the Second Law that the orders should not be obeyed if they conflict with the First Law already rule out hackers? Theoretically, it does, but hackers are often very creative and good at concealing what they are doing (Scroxton, 2023), and it may not be immediately obvious that the hacker is doing something harmful. To allow for algorithms that adapt (which is, in fact, a key characteristic of AI algorithms), we would have to reconsider the formulation of our Second Law. Drawing on Murphy and Woods (2009), who argue that robots must be built to fit the roles that individuals have, we suggest that the Second Law is revised:

(2) AI developers have a responsibility to see to it that an algorithm obey the orders given to it by *authorised* humans except where such orders would conflict with the First Law.

This revised Second Law takes into account which user the algorithm is obligated to obey and allows the disposal of orders exceeding the authorisation of the user. A difficulty is of course how to discriminate between authorised and unauthorised users, but to solve this primarily technical issue is a responsibility that our developer will have to take on.

The Third Law refers to protection of the algorithm. But for machine-learning algorithms that develop over time, what is it that should be protected? The developer's original blueprint of the algorithm as it was first put to work in the search engine? The algorithm as it is being developed by interaction with user data? And if so, at what point of time?

Considering the superior Second Law, which could be interpreted as the developer's responsibility not to design adaptive algorithms, it should be the original blueprint that the designer is responsible to protect. But as we indicate above, we do not want to rule out adaptation as such, but rather have a kind of adaptation that is not damaging for democracy. Our chosen interpretation of the Second Law assumes that the adaption is integral to the algorithm, and thereby should be protected. But regardless whether we want to protect the original or the adapted version of the algorithm, the algorithm needs to be protected from hackers, which the revised version of the Second Law should warrant.

There could, of course, be other causes of a destroyed algorithm than the act of hackers, which the Third Law directed at humans aims to take care of. For instance, suppose that the algorithm adapts to the extent that it not only exposes the individual to other world views than she already has but also to world views that are intrinsically damaging to humanity, say, world views that divide groups of people into superior and subordinate and are not worthy of equal treatment. Without doubt, that would be detrimental for democracy and for humanity. As Clarke (1993, 1994) reminds us, the laws should benefit all humans equally. Pasquale's (2017) idea on constraints on code evolution would be useful in this regard. Paraphrasing Pasquale, we propose that the developer is responsible to build in constraints to the algorithm such that to prevent outcomes that are damaging to democracy and harmful for humanity. This leads to the following specification of the Third Law:

(3) AI developers have a responsibility to see to it that an algorithm protects its own existence, *as it was intended by the developer*, as long as such protection does not conflict with the First or Second Law.

It appears that Asimov's Laws directed at humans, with some revisions, will be able to protect the functioning of democracy considering the context of technological development.

As it seems from this premature conjecture, the hierarchy between the First and the Zeroth Laws directed at humans solves the potential conflict between short-term individual interest and long-term societal interest. Hence, our developer needs to think more long-term than the robot did in 'Liar' and conclude that a well-functioning democracy is more important for individuals than to be protected from search results that go against their prevailing view. This will follow from prioritising the Zeroth Law before the First Law.

Regarding the Second Law directed at humans, there is a need of a specification of which humans that the algorithm should obey, which we achieve by making a distinction between authorised and unauthorised users. By that, we protect the algorithm from deliberate destruction by hackers.

To protect the algorithm from unintended destruction by its own adaptation, we add a specification to the Third Law that assigns responsibility to the developer to ensure that adaptation is constrained to the extent that the original intention of the algorithm is kept.

The revised version of Asimov's Laws of Robotics directed at humans taken together looks like this:

(1) AI developers have a responsibility to see to it that an algorithm may not injure a human being or, through inaction, allow a human being to come to harm.
(2) AI developers have a responsibility to see to it that an algorithm obeys the orders given to it by authorised humans except where such orders would conflict with the First Law.
(3) AI developers have a responsibility to see to it that an algorithm protects its own existence, as it was intended by the developer, as long as such protection does not conflict with the First or Second Law.

And the Zeroth Law:

(0) Developers have a responsibility to see to it that an algorithm may not injure humanity or, through inaction, allow humanity to come to harm.

Application on Imagined Future Scenarios

How would the revised version of Asimov's Laws of Robotics directed at humans work on future AI technology? In this section, we will tentatively apply the Laws to three imagined future scenarios in which AI has been developed well beyond today's level of sophistication. We take our departure in discussions on Super-intelligence and Artificial General Intelligence (AGI) (Bostrom, 2014; Russell, 2019; Tegmark, 2017) and prospects of conscious AI (Schneider, 2019), and imagine three significantly different and morally relevant directions of AI development. To try out how the revised Laws directed at humans apply in these cases, we adjust them to apply to the technology in question, but make no further changes.

Scenario 1: Super-intelligent AI that does *not* understand what it is doing, why it is doing it, or consequences of what it is doing.

With 'superintelligence' we follow Bostrom and refer to 'any intellect that greatly exceeds the cognitive performance of humans in virtually all domains of interest' (2014: 22). The super-intelligent artificial systems that Bostrom discusses in his famous paperclip example[1] does not have any kind of 'understanding', but a brute force rationality kind of intelligence and ability to act towards its goal to maximise paperclip production. Hence, the super-intelligent AI that we imagine here differs from today's AI systems in scope and efficiency, but not in kind, as it

---

[1]Bostrom's paperclip example refers to a thought experiment in which 'someone programs and switches on an AI that has the goal of producing paperclips. The AI is given the ability to learn, so that is can invent ways to achieve its goal better. As the AI is super-intelligent, if there is a way of turning something into paperclips, it will find it. [. . .] Soon the world will be inundated with paperclips' (Gans, 2018).

does not have any kind of moral awareness. What would that mean for the responsibility of our developer as it is stated in the Laws directed at humans?

The relation between the developer and the super-intelligent AI would be the same as in our contemporary hypothesised case. As this super-intelligent AI lacks moral agency, the principal responsibility of the developer will not be challenged. However, as this super-intelligent AI is far more efficient than contemporary AI and on practically all areas, the extent of the developer's responsibility to protect humanity and individual human beings will be dramatically larger. Still, this is not a principal, but a practical issue, yet a huge one.

Regarding the Third Law directed at humans, aimed at protecting the developer's intention for the super-intelligent AI, one practical issue that the developer will have to handle with particular care is how to ensure that these intentions (to protect humans and humanity) is formulated in a way that cannot be misunderstood by the super-intelligent AI. To align the goals of a potential future super-intelligent AI with human and societal values is a problem that is increasingly getting attention in the scholarly literature (Bostrom, 2014; Christian, 2020; Mechergui & Sreedharan, 2023; Russell, 2019; Søvik, 2022; Taylor et al., 2020, pp. 342–382). Besides technical questions on how values could be translated into machine code and incorporated into AI technology (Dignum, 2019), 'value alignment' involves both philosophical and societal aspects such as which values should be promoted and how to decide that (Savulesc et al., 2019; Smits et al., 2022; Hedlund, 2022).

For our future developer, we can only hope that these issues have been sufficiently solved. However, the point here is not to speculate whether that would be the case, but to illustrate the massive task ahead for our developer. While in principle, the developer's responsibility is the same as in the contemporary case, in practice, it magnifies with the scope of the intelligence of the AI.

Scenario 2: Super-intelligent AI that can act *as if* it understands what it is doing, why it is doing it, and consequences of what it is doing.

This super-intelligent AI shares the qualities with the super-intelligent AI in the first scenario, with the addition that it also has the ability to act as if it has moral agency. From a functional perspective, this is sometimes used as an argument for machine responsibility (Laukyte, 2017; Sullins, 2006). However, the fact that our future super-intelligent AI by interaction with the social world has learnt to imitate human interaction patterns does not lead us to the conclusion that it is a moral agent. Compelling biological, historical, and logical arguments speak against that (Gunkel, 2017; Hakli & Mäkelä, 2019; Sharkey, 2017). Hence, the fact that a super-intelligent AI can perform *as if* it is a moral agent does not make any morally relevant difference as compared to the first scenario, that is, the relation between our developer and this super-intelligent AI is qualitatively the same.

Like in the first scenario, the developer is responsible to see to it that the AI does not harm humanity or individual humans, and to do that, make sure that the AI continues to act according to the intentions of the developer. However, in addition to the demanding task of aligning the goals of the super-intelligent AI with human and societal values, the ability of this AI to act more humanlike will put even more arduous requirements on the developer. There is a risk that the behaviour of this super-intelligent AI might fool our developer in different ways.

For instance, the developer might believe that the super-intelligent AI in fact have moral agency and delegate responsibility to it in ways that she already does to other humans. (Remember the extremely heavy burden that by now rests on the shoulders of our developer.) Given that the work with value alignment that we introduced in the first scenario have been perfectly successful, this need not be a danger, but any loophole in that work is something that the super-intelligent AI would exploit if that would increase its chances to reach whatever goals it has been given. This could jeopardise compliance of all the Laws.

Another risk is that the capacity of this super-intelligent AI to imitate humans may make it capable to disobey human orders, but act in a way that makes the developer believe that it has in fact followed her orders. Obviously, in this setting, the developer does not have control in any strong sense over the AI. But could she have the weaker form of control in the sense of supervision? Considering that it is the developer who maintains, improves, and teaches the AI, she is arguably responsible for this in a way comparable to the responsibility of a parent for its child. However, if the AI's disobeying leads to harm for humanity or individual humans, then the developer has not taken appropriate responsibility according to the Third Law directed at humans, and failed in regard to the hierarchy between the Laws directed at humans.

Scenario 3: Super-intelligent AI that *understands* what it is doing, why it is doing it, and consequences of what it is doing.

A super-intelligent AI that understands what it is doing, why it is doing it, and consequences of what it is doing is a system that is conscious. There is no undisputed definition of 'consciousness', but sentiments, inner mental life, inner experiences, and subjective experience approximately captures what it is about (Schneider, 2019; Tegmark, 2017). As we indicated above, consciousness is also a requirement for moral agency. In the current discussion on conscious AI, two main positions are discernible, one claiming that consciousness is unique for biological beings, the other that consciousness is substrate independent (Schneider, 2019; Tegmark, 2017). In this imagined future scenario, we assume that substrate independent consciousness is possible, meaning that also silicon-based artefacts like our super-intelligent AI could be conscious. What would that mean for the responsibility of our developer to see to it that the super-intelligent AI does not harm humanity or human individuals?

As a moral agent, the super-intelligent AI has the capacity to be responsible. Would that somehow interfere with the responsibility of the developer? After all, it is the developer who has created this super-intelligent AI, and if it is conscious, that would be a result of this creation. But as we know already from contemporary times, an intrinsic feature of AI is its learning and adaptation, and the consciousness of this super-intelligent AI might be the result of such adaption and not of the intention of the developer. However, the Third Law entails that adaptation should be constrained to the extent that the intention of the developer is protected. This would probably rule out the option that the developer has not intended the super-intelligent AI to be conscious. Or would it? Perhaps superintelligence itself stands in the way for that option?

If the continuously increasing level of intelligence of the AI is accelerating exponentially, as Bostrom (2014) suggests, every new step of increasing intelligence is considerably larger than the previous one. Somewhere along this journey the developer might lose track. If consciousness emerges at a certain level of molecular complexity, as some 'techno-optimists' reason (Schneider, 2019), then consciousness could emerge even if that is not the intention of our developer. Then this scenario, with a conscious super-intelligent AI, is not the result of the developer's intention. The Third Law directed at humans never kicked in. Or, rather, the developer was unable to foresee the development and was thereby not able to take responsibility according to the Third Law directed at humans. For the same reason, she was not able to regard the hierarchy between the Laws directed at humans.

This tentative exposé of imagined future super-intelligent AI suggests that Asimov's Laws directed at humans would not be sufficient to protect humanity and individual humans in scenario 3, with a conscious super-intelligent AI, and, to some extent, in scenario 2, with a super-intelligent AI that can act as if it is conscious. The Laws would need to be complemented with some kind of precautionary principle. A precautionary principle is preemptive, tries to foresee the risks, and 'imposes some limits or outright bans on certain applications due to their potential risks' (Pesapane et al., 2018). However, the nature of AI systems, which adapt in an unforeseeable way according to their experiences and learning, makes this difficult, especially with regard to long-term development. For contemporary AI, the Laws directed at humans are more promising, at least for the particular case of protecting the functioning of democracy. This is, on the other hand, not a bad undertaking.

Conclusions

By taking Asimov's Laws of robotics as our point of departure, we tried out the option to direct them at humans and to incorporate the concept of 'responsibility' to do so. We applied our reformulation of the Laws directed at humans on a hypothesised real-case contemporary situation, namely, how recommender algorithms may damage the functioning of democracy.

This worked pretty well, given two specifying adjustments. In the Second Law, stating that a robot (or an algorithm) must obey human orders, a distinction between authorised and unauthorised humans was needed, and in the Third Law, we specified that the intention of the human creator of the robot (algorithm) must be protected. With these adjustments, we then applied the Laws to three imagined future scenarios, based on three significantly different and morally relevant directions of AI development: super-intelligent AI with no moral awareness, super-intelligent AI with the capacity to act as if is morally aware, and super-intelligent AI that is conscious and has moral awareness.

We found that the super-intelligent AI that lacks consciousness implies no principal problems for the human, albeit significantly larger responsibility in practical terms. As for the super-intelligent AI with the capability to act as if it has moral awareness, there is no morally relevant difference per se compared to the former case. However, the capability of this super-intelligence to act humanlike involves some risks, as the human developer might be deceived to believe that the super-intelligent AI is a moral agent, or that it is obeying the human when it in fact is not. If any of these outcomes materialises, then the Laws have not done their job. With the conscious super-intelligence, we come to the conclusion that the super-intelligence as such, due to is exponential development, makes the human unable to foresee the consequences, which makes her unable to take responsibility as stated in the Laws.

This tentative look into imagined future scenarios show that Asimov's Laws directed at humans seem to have a hard time in protecting humans and humanity when super-intelligent AI is or appear to be conscious, at least when the human is the only part responsible. Perhaps a viable way forward could be to hand over some of the responsibility to the AI? As it has moral awareness, or can act as if it has moral awareness, it would at least have the capacity to take responsibility. Could we imagine a future in which humans and the super-intelligent AI are responsible together? While this might look promising at a first sight, it would be a kind of collective responsibility, which gives rise to the problem of many hands. The problem of many hands refers to a situation when many agents contribute some part to an outcome that is dependent on each agent's contribution taken together. In such a situation, each of them is responsible for their own part, but none of the contributing actors is responsible for the entirety (Hedlund & Persson, 2024; van de Poel et al., 2015). This is a complicated situation when all involved agents are humans. When one of the agents is an artificial entity and many times more intelligent than the other, as in our imagined future scenarios, the involved agents stand in an asymmetrical relationship to each other, which would amplify the difficulties. Again, unless value alignment will be perfectly worked out, we cannot rule out the risk that the human will be deceived by the super-intelligent AI.

Finally, we would like to anticipate a possible objection to our approach to direct the Laws at humans. Considering the chosen examples of super-intelligent AI in our imagined future scenarios, it could be argued that we would not expect the Laws directed at humans to work. Given that our super-intelligence not only refers to intelligence that greatly exceeds the cognitive performance of humans but also to intelligence that is conscient, it is perhaps in the nature of things that humans would not have the capability to be responsible to protect humans and humanity from harm. The work with value alignment will have to start soon enough to be complete before we have a potential super-intelligence around us.

References

Abdullah, Y. I., Shuman, J. S., Shabsigh, R., Caplan, A., & Al-Aswad, L. A. (2021). Ethics of artificial intelligence in medicine and ophthalmology. *Asia-Pacific Journal of Ophthalmology*, *10*(3), 298.

Adam, B., & Groves, C. (2011). Futures tended: Care and future-oriented responsibility. *Bulletin of Science, Technology & Society, 31*(1), 17–27.

AI HLEG. (2019). *Ethics guidelines for trustworthy AI.* Brussels: European Commission: High-Level Group on AI. Report No. B-1049. https://ec.europa.eu/digital-single-market/en/news/ethics-guidelines-trustworthy-ai

Asimov, I. (1941). "Liar". *Astounding Science Fiction.* May issue.

Asimov, I. (1942). "Runaround". *Astounding Science Fiction.* March issue.

Asimov, I. (1985). *Robots and empire.* Doubleday.

Balkin, J. (2017). 2016 Sidley Austin distinguished lecture on big data law and policy: The three laws of robotics in the age of big data. *Ohio State Law Journal, 78*(5), 1217–1242.

Bostrom, N. (2014). *Superintelligence: Paths, dangers, strategies.* Oxford University Press.

Buruk, B., Ekmekci, P. E., & Arda, B. (2020). A critical perspective on guidelines for responsible and trustworthy artificial intelligence. *Medicine, Healthcare & Philosophy, 23*(3), 387–399.

Choi, Y. L., Choi, E. C., Chien, D. V., Tin, T. T., & Kim, J.-W. (2019). Making of South Korean robot ethics charter: Revised proposition in 2018. In *ICRES 2019: International Conference on Robot Ethics and Standards.* London, UK, 29–30 July 2019. https://doi.org/10.13180/icres.2019.29-30.07.004

Christian, B. (2020). *The alignment problem: Machine learning and human values.* W.W. Norton & Company.

Clarke, R. (1993). Asimov's laws of robotics: Implications for information technology, part 1. *Computer, 27*(1), 57–66.

Clarke, R. (1994). Asimovs laws of robotics: Implications for information technology, part 2. *Computer, 27*(2), 57–66.

Dignum, V. (2019). *Responsible artificial intelligence: How to develop and use AI in a responsible way.* Springer.

Duenser, A., & Douglas, D. M. (2023). *Who to trust, how and why? Untangling AI ethics principles, trustworthiness and trust.arXiv:2309.10318.*

EP. (2016a). European Parliament, Committee on Legal Affairs. *Draft report with recommendations to the Commission of civil Law Rules on Robotics.* 2015/2103(INL).

EP. (2016b). European Parliament, Legal Affairs. *European Civil Law Rules in Robotics.* PE 571.379.

EP. (2017). *European Parliament resolution "Civil Law Rules on Robotics".* P8_TA (2017)0051.

Fischer, J. M. (1982). Responsibility and control. *The Journal of Philosophy, 79*(1), 24–40.

Fischer, J. M., & Ravizza, M. (1998). *Responsibility and Control: A Theory of Moral Responsibility.* Cambridge University Press.

FitzPatrick, W. J. (2008). Moral responsibility and normative ignorance: Answering a new skeptical challenge. *Ethics, 118*(4), 589–614.

Fjeld, J., Achten, N., Hilligoss, H., Nagy, A. C., & Srikumar, M. (2020). *Principled artificial intelligence: Mapping consensus in ethical and rights-based approaches to principles for AI.* Berkman Klein Center for Internet & Society.

Gans, J. (2018). AI and the paperclip problem. *VoxEU* 10 June, 2018.

24 *Maria Hedlund and Erik Persson*

Gunkel, D. J. (2017). Mind the gap: Responsible robots and the problem of responsibility. *Ethics and Information Technology.* https://doi.org/10.1007/s10676-017-9428-2

Hagendorff, T. (2020). The ethics of AI ethics: An evaluation of guidelines. *AI & Society, 30,* 99–120. https://doi.org/10.1007/s11023-020-09517-8

Hakli, R., & Mäkelä, P. (2019). Moral responsibility of robots and hybrid agents. *The Monist, 102,* 259–275.

Hansson, S. O. (2012). Safety is an inherently inconsistent concept. *Safety Science, 50*(7), 1522–1527.

Hedlund, M. (2012). Epigenetic responsibility. *Medicine Studies, 3,* 171–183.

Hedlund, M. (2022). Distribution of forward-looking responsibility in the EU process on AI regulation. *Frontiers in Human Dynamics, 4, 703510.* https://doaj.org/article/1f56de7f628e405da7118151635101d3

Hedlund, M., & Persson, E. (2024). Expert responsibility in AI development. *AI & Society, 39*(2), 453–464. https://doi.org.10.1007/s00146-022-01498-9

Held, V. (1970). Can a random collection of individuals be morally responsible? In May, L., & Hoffman, S. (Eds.) *Five decades of debate in theoretical and applied ethics.* Rowman & Littlefield Publishers, Inc., pp. 89–100.

Himmelreich, J. & Köhler, S. (2022). Responsible AI through conceptual engineering. *Philosophy and Technology, 35,* 1. https://doi.org/10.1007/s13347-022-00542-2

Kaminka, G. A., Spokoini-Stern, R., Amir, Y., Agmon, N., & Bachelet, I. (2017). Molecular robots obeying Asimov's three laws of robotics. *Artificial Life, 23,* 343–350.

Kane, R. (2002). Responsibility, reactive attitudes and free will: Reflections on Wallace's theory. *Philosophy and Phenomenological Research, 64*(3), 693–698.

Laukyte, M. (2017). Artificial agents among us: Should we recognize them as agents proper? *Ethics and Information Technology, 19,* 1–17.

Matthias, A. (2004). The responsibility gap: Ascribing responsibility for the actions of learning automata. *Ethics and Information Technology, 6,* 175–183.

Mechergui, M., & Sreedharan, S. (2023). *Goal alignment: A human-aware account of value alignment problem. arXiv:230200813v2.*

Mill, J. S. (1859/2011). *On liberty.* Cambridge University Press.

Murphy, R. R., & Woods, D. D. (2009). Beyond Asimov: The three laws of responsible robotics. *IEEE Intelligent Systems, 24*(4), 14–20.

Nyholm, S. (2018). Attributing agency to automated systems: Reflections on human-robot collaborations and responsibility. *Science and Engineering Ethics, 24,* 1201–1219.

Pasquale, F. (2017). Toward fourth law of robotics: Preserving attribution, responsibility, and explainability in an algorithmic society. *Ohio State Law Journal, 78*(5), 1245–[iv].

Persson, E., & Hedlund, M. (2024). The Trolley problem and Isaac Asimov's first law of robotics. *Journal of Science Fiction and Philosophy, 7.* https://jsfphil.org/volume-7-2024-androids-vs-robots/asimovs-first-law-and-the-trolley-problem/

Pesapane, F., Volonté, C., Codari, M., & Sardanelli, F. (2018). Artificial intelligence as a medical device in radiology: Ethical and regulatory issues in Europe and the United States. *Insigths into Imagining, 9*(5), 745–753.

Russell, S. (2019). *Human compatible: Artificial Intelligence and the problem of control.* Viking.

Savulescu, J., Kahane, G., & Gyngell, C. (2019). From public preferences to ethical policy. *Nature Human Behaviour, 3*, 1241–1243.

Schneider, S. (2019). *Artificial you: AI and the future of your mind.* Princeton University Press.

Schurr, N., Varakantham, P., Bowring, E., Tambe, M., & Grosz, B. (2007). Applying laws of robotics to teams of humans and agents. In R. H. Bordoni, M. Dastani, & J. Dix (Eds.), *Programming multi-agent systems* (pp. 41–55). Springer.

Scroxton, A. (2023). Hackers: We won't let AI get the better of us. *Computer Weekly, 8*(1), 3–7.

Sharkey, A. (2017). Can robots be responsible moral agents? And why should we care? *Connection Science, 29*(3), 210–216.

Smits, M., Ludden, G., Peters, R., Bredie, S. J. H., van Goor, H., & Verbeek, P.-P. (2022). Values that matter: A new method to design and assess moral mediation of technology. *Design Issues, 38*(1), 39–54.

Sneddon, A. (2005). Moral responsibility: The difference of Strawson, and the difference it should make. *Ethical Theory & Moral Practice, 8*(3), 239–264.

Søvik, A. O. (2022). What overarching ethical principle shoulc a superintelligent AI follow? *AI & Society, 37*, 1505–1518.

Sullins, J. P. (2006). When is a robot a moral agent? *International Review of Information Ethics, 6*(12), 23–36.

Talbert, M. (2008). Blame and responsiveness to moral reasons: Are psychopaths blameworthy?. *Pacific Philosophical Quarterly, 89*(4), 516–535.

Taylor, J., Yudkowsky, E., LaVictoire, P., & Critch, A. (2020). Alignment for advanced machine learning systems, *Ethics in Artificial Intelligence* (pp. 342–382). Oxford University Press.

Tegmark, M. (2017). *Life 3.0: Being human in the age of Artificial Intelligence.* Penguin Random House.

Thompson, D. F. (1987). *Political ethics and public office.* Harvard University Press.

Torres, E., & Penman, W. (2021). An emerging AI mainstream: Deepening our comparisons of AI frameworks through rhetorical analysis. *AI & Society, 36*, 597–608.

Van de Poel, I., Royakkers, L., & Zwart, S. D. (Eds.) (2015), *Moral responsibility and the problem of many hands.* Routledge.

Chapter 2

Criminal Justice in the Age of AI: Addressing Bias in Predictive Algorithms Used by Courts

Rahulrajan Karthikeyan[a], Chieh Yi[b] and Moses Boudourides[b]

[a]Arizona State University, USA
[b]Northwestern University, USA

Abstract

As artificial intelligence and machine learning become increasingly integrated into daily life, both individuals and institutions are growing dependent on these technologies. However, it's crucial to acknowledge that such advancements can introduce potential flaws or vulnerabilities. A case in point is the investigation conducted by the non-profit organization ProPublica into the COMPAS (Correctional Offender Management Profiling for Alternative Sanctions) risk assessment tool – a tool widely used by US courts to assess the likelihood of a defendant reoffending. To address the issue of underlying biases, including racial biases, which can lead to inaccurate predictions and significant social harm, we are delving into the current literature on algorithmic bias in decision systems. We are also exploring the evolving considerations of fairness and accountability in machine learning. Specifically, within the realm of predictive policing algorithms employed in the criminal justice system, our focus is on recent studies aimed at mitigating biases in algorithmic decision-making. This involves reassessing recidivism rates and implementing adversarial debiasing in conjunction with fairness metrics.

Keywords: Bias; fairness; algorithmic decision; criminal justice; recidivism; machine learning; main body

---

The Ethics Gap in the Engineering of the Future, 27–50
Copyright © 2025 Rahulrajan Karthikeyan, Chieh Yi and Moses Boudourides
Published under exclusive licence by Emerald Publishing Limited
doi:10.1108/978-1-83797-635-520241003

Introduction

The concept of justice has long been regarded as a fundamental tenet of any society, wherein impartiality and objectivity are expected to guide judicial decision-making. However, the inherent biases of human judgement have historically undermined the efficacy of the criminal justice system, leading to disparities and inequalities in the administration of justice. That was why, in recent years, there was an increasing interest and research on a category of explorations of artificial intelligence (AI) and machine learning (ML) that are now referred to as 'algorithmic fairness' or 'fair ML' (Barocas & Selbst, 2016; Corbett-Davies et al., 2017; Cowgill & Tucker, 2019; Kleinberg et al., 2018). In particular, our aim here is to examine the use of AI and ML in addressing biases in the criminal justice system, highlighting their future.

The COMPAS algorithm serves as a notable example of potential biases inherent in AI-driven risk assessment tools used in the criminal justice system. As a matter of fact, COMPAS is a software tool, which uses various algorithms and data analysis techniques to assess a defendant's likelihood of reoffending (risk of recidivism) and it helps guide decisions related to criminal sentencing/supervision/ parole, etc. Typically, COMPAS is processing datasets of information about defendants' backgrounds, including several different attributes/features of individuals such as age, gender, criminal history, etc. Nevertheless, COMPAS has been found to exhibit significant racial and gender biases, with black and female defendants disproportionately affected. In 2016, an investigation by ProPublica revealed that the dataset was based on prior judgements made by biased judges, undermining the tool's objectivity and fairness. These findings underscore the need for greater scrutiny of AI and ML applications in the criminal justice system to ensure that they do not perpetuate or amplify existing biases.

This contribution aims to discuss existing fair and unbiased models of analysis in the case study of the COMPAS tool, given that the fairness and accuracy of algorithms applied to justice have frequently raised doubts among the public (Angwin et al., 2016). Before arguing about COMPAS, we are examining some common domain areas in which decision-making algorithms are habitually adopted. Additionally, we are conducting a literature review on existing/previous risk assessment tools.

Within the computational framework, while being engaged in replicating the existing analysis of the COMPAS recidivism algorithm (Larson et al., 2016), we have been employing a special toolkit AI Fairness 360 (AIF360) and the technique called adversarial debiasing. In principle, such an ML analysis involves two trained models: the first one is a Random Forest classifier and the other is the AIF360 debiasing model. Furthermore, a method called in-processing adversarial debiasing is often employed to make the model fair and unbiased even when some biases exist in the data. As it is done in ML processing, one needs to divide the data into two parts: a training set and a test set. This division helps the analyst to see how accurate and fair each model is and to be able to employ fairness metrics for the understanding of models that treat different groups of people fairly. Moreover, these metrics facilitate measuring the balance between fairness and accuracy. In this way, one can assess whether the model is making fair decisions for everyone while still being accurate in its predictions. By using these tools, one can build a better and more trustworthy model. Finally, to make the results even more reliable and understandable a technique called LIME is often employed.

Studies, such as ProPublica's scrutiny of the COMPAS tool, have important implications for the use of AI and ML in the criminal justice system. The approaches that we are discussing here show that one may use several techniques to overcome racial bias in the data so that one may create fair models that are still very accurate.

Domains of Algorithmic Decision Systems

In the rapidly advancing digital age, the integration of algorithmic decision systems (ADS) into various facets of our lives powered by ML and AI holds the promise of enhancing efficiency, objectivity and being cost-effective in decision-making processes across multiple domains. However, as the adoption of algorithmized processes continues to expand, an increasingly urgent question emerges: Can we genuinely trust algorithmic decision systems to make critical choices on our behalf? (Emspak, 2016) This is a concern, which is not merely theoretical. As has been already argued (Vartan, 2019), algorithms employed in ADS have the potential to perpetuate and even amplify society's deeply ingrained biases, contributing to widespread inequities. However, challenges involved in the practice of algorithmic decision-making systems extend far beyond this well-documented problem.

Incidentally, a recent commentary (Ananya, 2023) has focused on a previously underexplored aspect of their functioning in terms of how sensitive ADS are in minuscule variations of the way that humans annotate data used to train them. The latter issue delves into the complex landscape of ADS and its implications for various sectors from employment and healthcare to criminal justice and entertainment. It is an issue related to how these systems are already shaping who gets hired, who receives priority in medical care, how bail is determined, and even what content is consumed. While algorithmic systems promise to expedite decision-making, alleviate backlogs, and provide objective evaluations, there are plenty of news reports and many research findings uncovering the existence of certain alarming shortcomings in ADS, which can possibly have profound and long-lasting consequences in the lives of individuals affected by algorithmic judgements.

A recent study (Balagopalan et al., 2023) provides a particularly insightful perspective, which reveals how the nuances of human annotation during training can significantly influence algorithmic decisions. By examining how training in ML models is implemented to assess rule violations (such as dress code adherence or meal standards etc.), Balagopalan et al have uncovered some cases in research exhibiting a remarkable disparity in the outcomes of the employed models. Even when both versions of an algorithm were trained on the same rules, their judgements diverged, based on whether humans provided descriptive labels or directly evaluated the rule violations. This discrepancy, it turns out, reflects the fact that, when confronted with potential consequences of their decisions, humans may label data in multiple ways. Furthermore, the above discrepancy highlights the broader implications of relying on historical data for training algorithms. As ADS tend to infer the future based on past data, they may inadvertently perpetuate past biases and fail to adapt to evolving societal values. Thus, the following fundamental question emerges: Can algorithmic decision-making systems truly imagine a different future when their foundation is rooted in the past?

When decisions are solely based on patterns and principles derived from past data and predefined rules, ADS, unlike humans, are doomed to be inherently devoid of imaginative capacity and creativity. Although algorithmic decisions process historical data to make informed choices in the present, their ability to foresee a different alternative future is limited. Aspiring to make ADS consider a different future inadvertently necessitates adjustments to the employed underlying algorithms and data inputs, which are typically made available through human intervention. In other words, the stake of algorithmic fairness lies in a feasible rectification of ML algorithms that typically learn from an already biased past. Because ADS trained on questionable past data are prone to make unfair decisions possibly propagating discrimination in future cases (Pfeiffer et al., 2023).

Moreover, from an epistemic point of view, Holm (2023) is examining a related but neglected problem of algorithmic decision-making about how appropriate it is to allocate resources based on purely statistical evidence provided by algorithmic predictions. Stated formally, let us give the example when a malevolent event Y occurs, in such a way that all but one among N individuals might have played a role in Y. Thus, anyone among the N individuals has a probability of guilt equal to 0.99. Now, suppose the individual X is picked at random. Concluding that X was guilty, based only on the available purely statistical evidence, would be relying on a so-called actuarial inference. However, is an actuarial inference an adequate basis for punishing X? Most people would answer no because purely statistical evidence is epistemically deficient and inadequate to justify a verdict. This is not a criticism against the use of statistical generalizations and probabilistic inferences as such, but rather it is a criticism of the epistemic value of statistical evidence as justification for believing propositions of the form 'X is guilty of Y' (Holm, 2023, p. 28).

After having reviewed some of the general opportunities and risks related to the use of ADS and discussed the need to take into consideration other legal, ethical and social dimensions too, in what follows in this section, we are going to delve deeper into the findings of four different sectors of applications of ADS: (i) education, (ii) healthcare, (iii) employment and (iv) bank lending, leaving aside the case of criminal justice, which is going to be examined separately on its own in the next section. From shaping the allocation of resources and opportunities to influencing the delivery of services, algorithmic decision systems reach far and wide. Moreover, they are posing intricate challenges and ethical considerations. Surveying through the literature in these four sectors, we are aiming to gain a comprehensive understanding of many existing or potential benefits and, at the same time, to highlight the need for critical assessment in the deployment of ADS across diverse sectors of human lives.

(1) Education is a sector in which ADS has gained prominence, particularly in the admissions process intending to enhance efficiency and data-driven decision-making. However, algorithmic decision-making in education has also introduced the risk of bias, which can result in unfair outcomes, particularly for underrepresented groups. In what follows below, the significance of recognizing and addressing biases in educational institution admissions is underscored. Kizilcec and Lee (2022) offer the following valuable insights into bias and fairness issues within education decision algorithms.

As algorithmic bias in education might arise when ML models are trained on non-representative student data, it can result in models prone to making inaccurate predictions for specific student groups, such as the ones with minority backgrounds or students with disabilities. ADS also impacts the admissions process in educational institutions. For instance, low-income applicants who often have lower SAT scores and college GPAs may face higher rejection rates, and a model trained to predict student college success may unjustly predict failure for low-income students, solely because the employed training data might indicate lower graduation rates for this group (Kizilcec & Lee, 2022, pp. 174–102).

To mitigate algorithmic bias in education, several actions can be taken. First, it is crucial to ensure that the employed training data for ML models is representative of the entire student population. Achieving this involves data collection from diverse sources and oversampling data from underrepresented groups. This also entails assessing model performance across various student groups and identifying bias patterns. Once identified, steps can be taken to mitigate bias such as adjusting model parameters or employing post-processing techniques on model output (Kizilcec & Lee, 2022, pp. 174–102).

(2) Biomedical research and healthcare are some of the most important areas of applications of AI. In an important recent paper, Baumgartner et al. (2023) shed light on the challenges and opportunities that AI presents in these fields, emphasizing the importance of social science perspectives to ensure fairness, equity, and ethical use of AI.

In this perspective, one of the central concerns raised is the perpetuation of inequalities due to biased data and models. Biomedical AI systems heavily rely on data for their functioning and if the data used is biased or unrepresentative these systems can lead to inaccurate predictions and decisions. Demographic factors, such as race, gender, and socioeconomic status, play a significant role in shaping health outcomes and when not properly accounted for they can lead to health disparities. For instance, as Baumgartner et al. (2023, p. 3) argue, AI algorithms that are not trained on diverse datasets may struggle to accurately assess the needs of minority and marginalized groups.

Discrimination, exclusion, prejudice, and stereotyping are deeply rooted in social inequalities, which can also affect the algorithms used in healthcare AI (Baumgartner et al., 2023, p. 4). Consequently, there is always the risk that processing, labelling, and classification of data may inadvertently target certain populations while neglecting others. Thus, Baumgartner et al. (2023, p. 4) advocate for a justice-oriented design of AI algorithms, emphasizing the need for independent oversight and auditing by experts trained in science and technology studies (STS) to ensure that AI systems are fair, accountable and transparent.

Transparency in AI systems is another critical concern, in cases when many AI algorithms used in healthcare are complex and difficult to audit or explain (Baumgartner et al., 2023, p. 4). The authors of this paper conclude that the lack of transparency may leave both healthcare providers and patients in the dark about how decisions are made and this is something that may erode trust in AI systems. Therefore, the authors stress the importance of making AI systems more explainable and understandable for users.

As Beaulieu and Leonelli argue in their book (2021), the issue of distorted data and human errors in interpreting results is also a growing concern. It highlights the limitations of binary classifications particularly in gender and sex classifications and the need to recognize that neither sex nor gender is binary (Baumgartner et al., 2023, p. 4). Hence, overcoming these limitations requires a more nuanced approach to data collection and analysis.

Furthermore, within this realm, Chen et al. (2023, p. 2) insist on acknowledging the persistent challenge of discrimination based on race and gender during the evaluation of such critical factors as image acquisition, genetic variation, and intra-observer labelling variability in current clinical workflows. These discriminatory practices may inadvertently introduce bias into the decision-making processes. As one harnesses the transformative potential of ML and AI within the medical field, it is incumbent to shift the focus towards identifying and addressing design flaws inherent in AI algorithms. These flaws require meticulous handling and appropriate mitigation strategies.

Therefore, in a nutshell, many scholars emphasize the urgency of addressing fairness in AI for biomedical research and healthcare calling for an interdisciplinary collaboration (including perspectives from social sciences, race and gender studies, STS, and medical ethics) to develop and apply AI in a way that is equitable and beneficial for all (Baumgartner et al., 2023, p. 7).

(3) In the literature on algorithmic fairness in information systems, when it comes to issues of employment and processes of job hiring (Rieskamp et al., 2023), gender biases remain a persistently recurrent issue. For instance, Lavanchy (2018) has revealed a tendency among hiring managers to prefer male applicants over their female counterparts. Such a deeply ingrained bias raises serious concerns regarding fairness and equality in the recruitment process.

Nonetheless, the adoption of AI-based algorithms in Human Resource (HR) departments of organizations represents a significant advancement in streamlining and expediting the candidate evaluation process. In computerized HR management systems, resumes and applications are subject to a certain initial machine assessment before undergoing comprehensive manual review by human recruiters and reviewers. Notwithstanding, there are serious concerns among computer scientists and advocates of fair hiring practices on whether algorithmic hiring may make problematic decisions (Langenkamp et al., 2019).

That gender bias influences personnel selection processes has been already documented in numerous field experiments (for the Spanish case study, Martínez et al., 2021), which examine the causes resulting in gender inequalities and discrimination in job access, hiring decisions, selection of leaders, salaries, etc. (World Bank Group Study, 2020).

Thus, recent research and systematic reviews in the field of HR recruitment and development have shed light on the challenges, causes and consequences of algorithmic decision-making. Algorithms have been shown to produce discriminatory or biased outcomes when trained on inaccurate or invasive exploitation of sensitive personal data (Kim, 2017), on societally prejudiced cases (Barocas & Selbst, 2016), or input data instrumental in representational harm (Suresh & Guttag, 2021, pp. 1–9). As a matter of fact, all these issues underscore the vulnerability of algorithms to replicate biased decisions if their training data is inherently flawed or biased.

Moreover, the employment of Natural Language Processing (NLP) and ML tools in the evaluation of such data as resumes or interview transcripts has demonstrated biases against women and individuals with disabilities (Engler, 2021). Speech recognition models have exhibited clear biases against African Americans and have struggled with dialectical and regional variations in speech (Koenecke et al., 2020). The use of commercial AI facial analysis has revealed disparities across skin colour and raises significant concerns, particularly for individuals with disabilities (Engler, 2019). Algorithms employed in the advertisement of job postings have been found to unintentionally lead to biased outcomes, including bias against young women seeking STEM jobs, as well as ageism against older candidates (Lambrecht & Tucker, 2019). The cumulative impact of all these algorithmic biases on hiring decisions is particularly troubling since the accumulation of small biases within algorithms can create larger structural effects.

Ensuring the fairness of AI systems in job hiring is paramount, especially considering that job applicants can be directly rejected by certain inadequate algorithmic systems. Applicants' data attributes like sex, ethnicity, or age may be responsible for biases harming and potentially disfavouring certain groups of individuals (Chakraborty et al., 2021). Biases in AI-driven hiring processes can stem from the so-called cognitive biases, such as the microeconomic home bias, similarity bias in recruitment, or some other stereotypes that can lead recruiters to favour specific applicants over others (Rieskamp et al., 2023). Measurement bias relates to general errors in data collection, while representation bias occurs when data fails to accurately depict the relevant population resulting in an underrepresentation of certain groups (Suresh & Guttag, 2021, pp. 4–5). Algorithms can also exhibit bias due to specific optimization functions or the use of biased estimators (Baeza-Yates, 2018). Historical biases inherited through the transition to AI-based algorithms can further exacerbate these issues (Australian Human Rights Commission, 2020).

Efforts to mitigate gender and ethnicity bias in AI-based hiring processes can be implemented throughout various stages of data processing. According to Rieskamp et al. (2023, p. 223), the following checklist of computational actions might ensure algorithmic fairness:

- Pre-processing involves costly efforts to enhance accuracy and denoise protected attributes.
- In-process tweaks introduce adversarial networks to mitigate gender and ethnicity bias with evidence suggesting the existence of increased fairness, albeit at the cost of reducing model accuracy.
- Post-processing approaches aim to rectify bias by reranking processed data to include candidates from protected groups. For instance, in the case of gender bias, high-ranked candidates can be combined with a gender distribution over qualified candidates to achieve a new and fair ranking.
- Feature selection methods have also been employed to remove features containing information about candidates' gender, resulting in a substantial reduction in bias.

In a nutshell, the previous scrutiny of the existing literature suggests that the prevalence of gender bias in hiring processes necessitates a concerted effort to leverage AI technologies to eliminate biases and promote fairness and equity in recruitment. Organizations must adopt strategies and techniques at various stages of the hiring process to ensure that AI-based algorithms treat all candidates equally regardless of their demographic (or other) personal attributes. By addressing these biases, one can move closer to achieving fair and equitable hiring practices that are solely based on candidates' qualifications and abilities.

(4) In the field of bank lending and mortgage credit, racial bias casts a long shadow over bank lending, resulting in unequal treatment among different ethnicities (Counts, 2018). This bias becomes evident when examining the disparities in mortgage credit approval rates for various racial groups. For instance, individuals of European descent often find themselves with a higher likelihood of mortgage approval compared to their counterparts of Asian descent (Martinez & Kirchner, 2021).

Bhutta et al. (2022) have unveiled troubling disparities in mortgage approval recommendations and processes. They discovered that Black and Hispanic applicants tend to face significantly lower mortgage credit scores compared to their white counterparts. Additionally, their research revealed that Black and Hispanic applicants are less likely to secure loans from government-backed sources. This disheartening pattern is further underscored by a persistent

Criminal Justice in the Age of AI **35**

Black-white denial gap of two percentage points alongside residual gaps for Hispanic and Asian applicants of approximately one percentage point. These residual disparities are aptly referred to as excess denials which further highlights the existence of racial bias within mortgage lending practices. While algorithms have been increasingly employed in the lending process, it is crucial to recognize that they too must adhere to fair lending regulations. This means that these algorithms cannot factor in race or ethnicity or proxies such as neighbourhood location or ZIP code. While the integration of algorithms into lending processes has the potential to mitigate discrimination compared to face-to-face lenders it falls short of completely eliminating discrimination in loan pricing (Bhutta et al., 2022, p. 7).

Bartlett et al. (2019) have shed additional light on this issue. They have estimated that lenders exhibit discriminatory practices against Latinx and African-American applicants, resulting in higher rejection rates and the imposition of higher percentage interest rates for purchase mortgages. This alarming revelation underscores the systemic nature of racial bias in the way that ADS operates within the lending industry. Furthermore, the authors have argued that, even with the advent of algorithmic lending, discrimination remains a persistent issue, especially concerning loan pricing (Bartlett et al., 2019, pp. 2–5). Lenders must adhere to a stringent set of criteria to ensure that their lending practices are fair and unbiased. This includes demonstrating that variables such as high school attendance were correlated with historical data related to fundamental lifecycle variables such as income growth and they did not predict protected characteristics after orthogonalizing them to the lifecycle variables (Bartlett et al., 2019, p. 22).

After having described how algorithmically-driven decision-making systems are employed in four representative sectors and having discussed certain problems caused by the inherent biasing that algorithms used in these systems may display, we are moving to our main focus, which is ADS in criminal justice. For this purpose, we are starting, in the next section, with a discussion of the questionable role that algorithms of recidivism prediction play in certain frequently utilized assessment tools in the criminal justice system.

Recidivism Risk Assessment Instruments

Recidivism risk assessment instruments serve as pivotal tools in the criminal justice system aiding judges/jury in predicting the likelihood of defendants reoffending after their initial encounter with the legal system. In this respect, Fazel et al. (2022) have identified the following 11 widely used risk assessment instruments in criminal sentencing.

(1) One of the first instruments of this sort was the COMPAS tool, which was developed by the private company Northpointe Inc. (now Equivant), where the acronym COMPAS stands for 'Correctional Offender Management Profiling for Alternative Sanctions.' This is a standard instrument used in the US criminal justice system to forecast a defendant's likelihood of

reoffending (recidivism) and it has relied upon the sentencing decisions made by certain judges. Moreover, COMPAS has attracted the groundbreaking investigative journalism work of the non-profit organization ProPublica (Barenstein, 2019). Since the COMPAS tool is going to be discussed separately in the next section, we are now examining the remaining 10 tools in more detail.

(2) Douglas et al. (2014) have introduced the assessment tool of Historical Clinical Risk Management-20 (HCR-20), which is a comprehensive instrument that amalgamates historical clinical and risk management factors to gauge the potential for violent or aggressive behaviour. This is a multidimensional approach acknowledging the significance of a holistic comprehension of an offender's background and circumstances in predicting the risk of recidivism. HCR-20 evaluates variables such as past criminal activities, mental health and contextual triggers. Thus, it offers a nuanced perspective on a complex interplay among factors contributing to tendencies of reoffending.

(3) Latessa et al. (2013) delve into the Indiana Risk Assessment System (IRAS), which is made up of five instruments designed to assess recidivism risk and guide intervention strategies. This tool takes a comprehensive approach by considering various elements including criminal history, behaviour and the social environment. Notably, the IRAS incorporates dynamic factors such as an individual's participation in treatment programs, thus, recognizing the offender's potential for positive change in evolving circumstances.

(4) Andrews et al. (2004) have introduced an essential recidivism risk assessment instrument, called Level of Service/Case Management Inventory (LS/CMI). This instrument takes into account a range of dynamic and static factors, thus, providing a holistic perspective on an offender's risk profile. Its comprehensive nature allows it to evaluate aspects such as criminal history, family and marital relationships, employment, substance abuse, and offender's attitude and orientation. LS/CMI is particularly notable for its emphasis on case management and intervention planning. Indeed, it goes beyond risk assessment by guiding professionals in developing tailored case management strategies based on the identified risk factors. In this instrument, Andrews et al. (2004) developed the Risk-Need-Responsivity (RNR) model emphasizing the importance of addressing an individual's criminogenic needs to reduce the likelihood of reoffending effectively. In this way, LS/CMI becomes a valuable asset in promoting various interventions (mostly, cognitive-behavioural ones), which may differ in the corresponding degree of effectiveness to reduce recidivism. Furthermore, the model of LS/CMI may encompass various attributes of offenders, such as gender, cognitive capacities, motivations, etc.

(5) Andrews and Bonta (1995) have introduced the actuarial assessment tool, Level of Service Inventory-Revised (LSI-R). Like LS/CMI, LSI-R takes a comprehensive approach, taking into account a range of dynamic and static factors providing again a holistic perspective on an offender's risk profile. It acknowledges that an individual's circumstances and behaviours can

change over time. Therefore, LSI-R incorporates ongoing assessment and reassessment ensuring that interventions remain relevant and effective in reducing the risk of recidivism. Like LS/CMI, its approach aligns with the Risk-Need-Responsivity (RNR) model and it also may develop tailored case management strategies on the identified risk factors.

(6) Garrett et al. (2019) established the Nonviolent Risk Assessment (NVRA), which is a valuable addition to the field of risk assessment tools within the criminal justice system. This tool is specifically designed to provide a targeted approach in the assessment of risk factors associated with nonviolent or low-level offending conduct (nonjail alternatives). In this context, nonviolent behaviour is understood to include factors, such as mental health, substance abuse, family, social considerations, and individual behaviour patterns. While NVRA excels in assessing nonviolent risks, it may require complementary tools or interventions for case management and rehabilitation.

(7) Howard (2006) has developed the Offender Assessment System (OASys) used by the prison and probation services in England and Wales. This is an assessment tool that evaluates both the risk of harm and it identifies – classifies offending-related needs, including basic personality characteristics and cognitive behavioural problems needs of offenders. OASys focuses on analysing dynamic factors such as attitudes, employment status, and substance abuse, which can significantly influence an individual's likelihood of reoffending. It underscores the importance of tailored interventions that address specific needs, which ultimately aim to reduce the risk of reoffending by addressing underlying factors contributing to criminal behaviour.

(8) Latessa, Lemke, Makarios, and Smith (2010) have discussed the Ohio Risk Assessment System (ORAS), which is a state-wide comprehensive instrument tailored to assess recidivism risk of adult probationers. This system offers a structured framework to guide supervision strategies, thus, ensuring that resources are allocated effectively to mitigate the risk of reoffending. ORAS elaborates on the notion that interventions should be targeted based on an individual's risk profile aligning with the broader principle of risk-needs-responsivity.

(9) The Canadian forensic psychologist Robert D. Hare (1993) has elaborated the Psychopathy Checklist-Revised (PCL-R), an essential assessment tool in the field of forensic psychology and criminology. It is a tool specifically designed to assess individuals diagnosed with psychopathy vs ones diagnosed with antisocial personality disorder. PCL-R is based on a detailed checklist of clinical cases of psychopathy (Huchzermeier et al., 2007). It is a multidimensional approach to assess various factors including interpersonal, affective, and lifestyle dimensions to determine the presence of severity of psychopathy in individuals. What sets PCL-R apart is its utility in assessing individuals who may pose a higher risk of engaging in serious criminal behaviour due to psychopathy. The PCL-R yields a total score that ranges from 0 to 40 with higher scores signifying a heightened level of

psychopathy (Hare, 1993). In professional assessment, a total score within the range of 10 to 19 on the PCL-R denotes a mild degree of psychopathy, while a score falling between 20 and 29 indicates a moderate level and a score between 30 and 40 signifies severe psychopathic. This finely-tuned scoring system aids the precise categorization of individuals based on the severity of their psychopathy and provides valuable insights for risk assessment and intervention strategies (Scarlet, 2011).

(10) Lowenkamp et al. (2013) have developed the Post Conviction Risk Assessment (PCRA) tool, which is a valuable instrument in the hands of the federal supervision system in the USA. This instrument is designed to assess the risk of recidivism in individuals, who have already been convicted and sentenced and it plays a role in informing decisions related to parole, re-entry programs and supervision. It takes into account a wide range of factors including an individual's criminal history, institutional behaviour, and other dynamic variables that may influence the likelihood of reoffending. This approach seamlessly aligns with the Risk-Need-Responsivity (RNR) model that emphasizes the importance of addressing an individual's criminogenic needs for successful reintegration into society.

(11) Another actuarial tool for assessing recidivism risk was introduced by Helmus et al. (2011): the Static-99 Revised (Static-99(R)), an updated iteration of the original Static-99, which retains the core principles of its predecessor. Static-99(R) addresses a critical need in risk assessment by evaluating unalterable factors contributing to an individual's risk of recidivism, specifically in cases of individuals with prior sexual offence convictions, and it incorporates refinements and adjustments informed through empirical research and clinical insights (Melton et al., 2018). As it is described in the manual of the third published version of the Static-99 (Phenix et al., 2016), what sets the Static-99(R) apart is its focus on evaluating static risk factors, such as an individual's criminal history, age at release and victim gender. By emphasizing these unchangeable variables, Static-99(R) provides a valuable perspective on an individual's long-term risk profile. This approach complements other tools that assess dynamic risk factors thus enabling a comprehensive evaluation of an individual's overall risk. Overall, the significance of the Static-99(R) lies in its potential to inform decisions related to sentencing, parole, and post-conviction supervision, thereby contributing to the safety and fairness of the criminal justice system (Phenix et al., 2016, p. 30 and p. 35). In terms of offenders' relative risk for sexual recidivism, the Static-99(R) utilizes readily available demographic and criminal history information known to correlate with sexual recidivism among adult male sex offenders. In this way, such a categorization facilitates law enforcement in developing tailored supervision strategies for each individual (Phenix et al., 2016, p. 10, p. 13, and p. 23). The main strengths of the Static-99(R) include setting up several explicit rules and following strictly the relative objectivity of a scoring system. The total risk score is determined through specific risk factors outlined in a scoring table (given as the tally sheet on page 99 of Phenix et al., 2016) that

provides a ranked assessment of each sexual offender. Hence, by adhering to the coding manual and the established rules, evaluators can achieve a more objective and less biased decision-making process and, eventually, contribute to enhancing the tool's reliability.

In a nutshell, concluding the discussion of the above 11 assessment instruments, one may remark that they all provide valuable insights into the realm of recidivism risk assessment. Overall, these tools offer multifaceted perspectives on risk, encompassing historical, clinical, dynamic, and comprehensive factors. While each instrument approaches risk assessment both from a distinct angle and collectively, these 11 tools underscore the importance of holistic evaluations and tailored interventions in promoting a fairer and more effective criminal justice system.

ProPublica's Study COMPAS

Going back to the first of the 11 instruments, COMPAS, let us mention that in 2016, ProPublica, a non-profit news organization specializing in investigative journalism, investigated the use, fairness, and accuracy of the COMPAS algorithm (Larson et al., 2016). ProPublica's investigation revealed that the tool was racially biased against African-American defendants, resulting in harsher and more frequent sentences. As a matter of fact, in their investigation of the COMPAS algorithm's predictions for over 10,000 defendants in Broward County, Florida, ProPublica discovered troubling evidence of racial bias (Angwin et al., 2016, p. 2). ProPublica's analysis revealed that the algorithm disproportionately labelled black defendants as high-risk for reoffending, even if they did not actually re-offend, while white defendants were more likely to be labelled as low-risk, even when they did go on to commit further crimes (Angwin et al., 2016, p. 2).

Moreover, the examination of the COMPAS instrument by ProPublica uncovered that the accuracy of the algorithm was significantly lower for female defendants, particularly for black women who were disproportionately misclassified as high-risk (Larson et al., 2016, p. 5). These findings reveal the gender and racial biases inherent in the algorithm and underscore the urgent need to identify and address such biases in ML tools employed in the criminal justice system. In particular, ProPublica has shown that 45% of black defendants were misclassified as high-risk compared to only 23% of white defendants. In addition, white defendants who reoffended were misclassified as safe 48% of the time compared to only 28% for black defendants (Angwin et al., 2016, p. 7). These differences are even more significant when controlling for relevant criminal history, age, and gender, with black defendants being 45% more likely to be given a higher risk score than their white counterparts (Angwin et al., 2016, p. 7).

ProPublica's investigation prompted significant attention and generated discussions about the implementation of algorithms in the criminal justice system (Washington, 2019, pp. 11–12). Moreover, it illuminated the importance of transparency and accountability when it comes to decision-making based on algorithms, particularly concerning sensitive issues such as criminal justice. The results of the investigation raised concerns regarding the fairness of algorithms and the potential for unintended bias and discrimination, emphasizing the need for further research and development of more equitable ML practices.

Nonetheless, as has been already noted (Barenstein, 2019), there exists a significant flaw in the COMPAS system. Rather than mitigating the systemic racism inherent in the American criminal justice system, it served to perpetuate it, since the system was trained on data from biased judicial decisions, it effectively reinforced existing biases (Park, 2019). Judges who sought to remain impartial relied on COMPAS's predictions, but in doing so unwittingly replicated the flawed decision-making of their biased peers, something which amplified the issue, leading to even greater disparities in the system (Larson et al., 2016).

Thus, the biased data used to train the COMPAS model resulted in flawed decision-making in the criminal justice system. For instance, black defendants charged with drug possession were more likely to be labelled as high-risk compared to white defendants with a more serious criminal history (Angwin et al., 2016, p. 4). Additionally, a black female with only juvenile misdemeanours and no new crimes after two years was rated as a higher risk than a seasoned criminal with prior convictions and a lower risk of reoffending (Angwin et al., 2016, pp. 1–2). The use of COMPAS in the criminal justice system highlights the grave danger of turning a blind eye towards the biased data that might have been used to train the model, regardless of the good intentions behind its development.

ProPublica's rigorous investigation of COMPAS revealed the grave dangers of biased data and underscored the pressing need for ethical considerations in the development and deployment of ML models in critical decision-making processes. As a result of ProPublica's reporting, COMPAS has been widely contested (Bao et al., 2021, p. 11; Goel et al., 2021; Taylor A., 2020), signalling a significant victory in the fight against systemic racism in the criminal justice system. This serves as a cautionary tale that, to foster fair and equitable outcomes, it is imperative to scrutinize and rectify the biases that may be embedded in the data used to train ML models.

Algorithmic Biases and Debiasing

The COMPAS algorithm's issues of bias and fairness in algorithmic decision-making have been highlighted by numerous studies. In particular, Dressel and Farid (2018) revealed that African-American defendants were twice as likely to be labelled as high-risk compared to white defendants with similar backgrounds. Meanwhile, Chouldechova's (2017) study showed that the use of COMPAS led to higher rates of false positives for African-American defendants, resulting in longer sentences and increased incarceration rates. Additionally, research by Angwin et al. (2016) and Larson et al. (2016) demonstrated that the use of risk assessment tools like COMPAS can lead to racial disparities in sentencing, with African-American defendants receiving harsher sentences than their white counterparts for similar offences. ProPublica's ground-breaking investigation into the COMPAS algorithm adds to this body of work, underscoring the need for ethical considerations in the development and deployment of ML models in critical decision-making processes.

In summary, these studies highlight the pressing need for algorithmic decision-making tools to be developed and assessed with fairness and accountability in mind. ProPublica's investigation into the COMPAS algorithm has prompted a broader dialogue regarding the use of algorithmic decision-making in the criminal justice system, emphasizing the crucial importance of transparency and accountability in this field.

Adversarial debiasing refers to a set of techniques used to mitigate the effects of biases in a dataset during the construction of a fair ML model. In-processing adversarial debiasing, a prominent method within this family of techniques involves training the primary model to predict the target variable, while simultaneously training an adversary to predict the biased attribute of the input data. The goal of this method is to avoid a correlation between the protected attribute and the primary model's decision-making, which can lead to biased outcomes (Yang et al., 2023).

In the process of adversarial debiasing, the primary model is trained to maximize accuracy while confusing the adversary. This approach enables the model to learn how to focus on relevant input features while disregarding biased attributes. In addition, the model is designed not to base decisions on features that overly correlate with the protected attribute, thereby minimizing the potential for perpetuating biased outcomes in the decision-making process. For example, ZIP codes have historically been used as a surrogate for race due to the history of segregation in the United States, and hence, adversarial debiasing applied to a model for removing racial bias can also help mitigate ZIP code bias.

Adversarial debiasing is a flexible and effective tool for addressing bias in datasets, particularly in cases where correlations between the protected attribute and other features in the data could potentially perpetuate biased outcomes. By removing this correlation, adversarial debiasing can produce a model that is free from bias and can be used in critical decision-making processes without fear of perpetuating systemic bias.

Correa et al. (2023) have in particular addressed biases in the context of medical image analysis. They proposed a two-step framework that incorporated adversarial debiasing and partial learning. In the first step, they used a Generative Adversarial Network (GAN) to achieve statistical learning of a latent representation of medical images that is less influenced by biased factors. In the second step, they trained separate models on a debiased subset and a biased subset to identify and correct biases in the latter.

In another influential paper, Zhang et al. (2018) explored the use of adversarial learning to address bias in ML models. They proposed an adversarial learning framework that consisted of a classifier and an adversary. The classifier aimed to predict the target variable accurately while the adversary strived to identify the sensitive attributes used by the classifier to make predictions. By simultaneously training the classifier and the adversary, they achieved a model that is both accurate and unbiased. Their work has highlighted the effectiveness of adversarial learning in debiasing ML models.

The Problems With the COMPAS Algorithm

The COMPAS algorithm originally processed a dataset containing information on 11,757 criminal defendants, who were assessed in Broward County, Florida, between 2013 and 2014 (Blomberg et al., 2010). The dataset included demographic information, such as age, gender, and race, as well as data on criminal history, socioeconomic status, and neighbourhood characteristics. Additionally, the dataset incorporated the predicted scores for recidivism for each defendant, which was a key variable used in the analysis of the dataset.

Nonetheless, the use of the COMPAS system has been widely criticized for its bias in predictions, particularly in its tendency to misclassify black defendants as higher risk and white defendants as lower risk. This bias is believed to be partially attributed to the use of historical data on criminal offenders that reflect societal biases and discrimination, as well as the inclusion of variables that are correlated with race (Israni, 2017). In general, including demographic information such as race in the dataset is not inherently problematic, but rather the issue arises when such information is used to make biased predictions as in the case of the COMPAS algorithm.

Although it is important to note that the COMPAS algorithm's specifics are not available to the public, we have attempted to reprocess the COMPAS analysis in a course project (Karthikeyan et al., 2023). After applying a range of data clean-up techniques to ensure the data was consistent and reliable, we were left with approximately 7,000 criminal defendant records. Thus, the dataset included demographic details, such as age, gender, and race, alongside more relevant information such as prior criminal history and the degree of the criminal charge that a defendant was facing. Subsequently, we ran three primary experiments, two with the baseline Random Forest model (Hastie et al., 2009, pp. 587–604) and one with an adversarial debiasing model. In particular, we have been also using two ML tools:

- LIME (being the acronym for 'Local Interpretable Model-agnostic Explanations'), a technique that can trustworthily explain the predictions of any classifier or regressor by approximating it locally with an interpretable model (Ribeiro et al., 2016).
- AIF360 (AI Fairness 360) is a toolkit comprising a set of open-source tools developed by IBM (including fairness metrics for datasets and models together with the corresponding explanations and algorithms). According to Bellamy et al. (2019), 'the main objectives of this toolkit are to help facilitate the transition of fairness research algorithms for use in an industrial setting and to provide a common framework for fairness researchers to share and evaluate algorithms.'

In the first experiment (Random Forest model with LIME), the objective was to find the baseline model accuracy with the biased dataset and identify the primary relevant data attributes. For this purpose, a minimalistic 100-estimator Random Forest model was trained to predict real two-year recidivism. Eighty percent of the

Criminal Justice in the Age of AI   **43**

pruned dataset was used for training, while the remaining 20% was used for validation. The result found the model's accuracy at 76.99%, where the five primary features used for prediction were: (i) the date defendants left custody (if they were recently in custody), (ii) the date that they were screened, (iii) how long since they were most recently put into custody, (iv) the date of their most recent arrest and (v) how many days between their screening date and their arrest date.

In the second experiment (adversarial debiasing with LIME), the same selection and division of training and test data were utilized and a 200-hidden-unit network was constructed using the AI Fairness 360 library (AIF360) to predict two-year recidivism while treating both race and sex as protected attributes that should not be recognizable given the other features and the final result. The found model's accuracy was 76.09%, a very slight reduction from the Random Forest implementation. The most relevant primary features for prediction were now: cases (i), (iii), (iv) of the first experiment, (vi) the date defendants most recently left jail and (vii) the defendants' juvenile felony count.

In the third experiment (Random Forest with fairness metric), the objective was to gain information about the fairness metrics specifically regarding race. Following a similar methodology, the observed demographic parity was found to be approximately 0.149 while equalized odds were approximately 0.106.

Based on the above experiments, we found that both the baseline Random Forest model and the adversarial debiasing model performed similarly in predicting two-year recidivism. However, there was an incredibly minimal accuracy drop and a much higher performance increase than what was expected from the debiased model.

In terms of feature importance, the Random Forest model relied on a peculiar collection of seemingly unrelated dates, while the adversarial debiasing model found that defendants' recent arrest record and their criminal history were crucial for predictions. Interestingly, we found that neither made a significant priority of race. Because the Random Forest regression model was trained to predict real-world recidivism rates and not to recreate the original COMPAS scores, it did not fully replicate the mistakes of its predecessor due to the increased knowledge of the dataset. Despite this, the nonsensical priorities of the Random Forest model suggest that there might remain an issue with the data. The fact that these trends disappear when using adversarial debiasing indicates that they were trends having significant correlations with race, even if that relationship was unclear from a pragmatic view.

Overall, our findings suggest that the adversarial debiasing model can achieve similar accuracy to the baseline Random Forest model while mitigating bias. The use of fairness metrics can also be useful in understanding the impact of sensitive attributes on model predictions. These results have important implications for developing fairer and more accurate ML models for predicting two-year recidivism.

Conclusions and Further Studies

As AI and ML are entering human lives, many applications have leveraged the benefits of the technology. However, as mentioned previously, along with the influence of AI and ML becoming stronger, scientists are questioning their reliability and fairness. Especially, an algorithmic decision system like COMPAS that can impact not only the criminals but also other potential victims has come to public attention. In our empirical work (Karthikeyan et al., 2023), we aimed to analyse and suggest improvements to the COMPAS algorithm from different aspects through three experiments.

In the study of our course project (Karthikeyan et al., 2023), we compared the performance of a baseline Random Forest model with an adversarial debiasing model, which was predicting recidivism. The results showed that both models performed similarly in accuracy, with the Random Forest model having a slightly higher accuracy than the adversarial debiasing model. However, the adversarial debiasing model identified more relevant features for prediction, including recent arrest records and criminal history, while the Random Forest model prioritized a bizarre collection of seemingly unrelated dates that bore unexpected correlations with the defendant's race. In addition, the fairness metrics that we have employed showed relatively good values for demographic parity and equalized odds in the Random Forest model, although there is still room for improvement to achieve a fairer system using techniques such as the often-mentioned adversarial debiasing method. These findings have important implications for the development of fairer and more accurate ML models that will predict recidivism. The adversarial debiasing model demonstrated that it is possible to achieve similar accuracy while mitigating bias, and the use of fairness metrics can be useful in understanding the impact of sensitive attributes on model predictions.

Future computational studies that we intend to undertake are going to continue to explore ways to improve the fairness and accuracy of ML models in the criminal justice system, paying attention to the human bias that often makes its way into seemingly impartial statistics. One direction that we intend to follow is that of the reduction of bias in the data using methods like Gradient Penalty, Disparate Impact Remover, and SMOTE.

- Gradient Penalty has been used in the context of improving the performance of the Wasserstein Generative Adversarial Networks (WGANs) (Arjovsky et al., 2017), which generate only poor samples or fail to converge. Gulrajani et al. (2017) found that using a certain weight clipping in WGANs to enforce a Lipschitz constraint on the discriminator was the source of many problems leading to undesired behaviour. Thus, by penalizing the norm of the gradient of the discriminator, Gulrajani et al. proposed the Gradient Penalty (WGAN-GP), which does not suffer from the same problems.
- Disparate Impact Remover (Feldman et al., 2015) is a pre-processing technique that by editing feature values increases group fairness while preserving rank-ordering within groups. The motivation behind the work of Feldman et al. was that, in the legal system of the USA, unintentional bias is often encoded via the so-called disparate impact, which occurs when the outcomes of a selection process for different groups (e.g., over ethnicity, gender, etc.) turn out to be disparate. In their 2015 publication, Feldman et al. have managed to

link disparate impact to an underrated measure of classification accuracy and proposed a test for disparate impact based on how well a class can be predicted from existing empirical attributes in data. In this way, they developed a methodology on how data might be made unbiased.

- The Synthetic Minority Over-sampling Technique (SMOTE) is one of the very often used methods to mitigate bias in AI models, which happen to result in unfair decisions. Chawla et al. (2002), who established the SMOTE method, were concerned with the construction of classifiers from imbalanced datasets (a dataset is imbalanced if the classification categories are not approximately equally represented). They found that a combination of their method of over-sampling the minority class and under-sampling the majority class can achieve better classifier performance than only under-sampling the majority class. Recently, the mathematical justification of SMOTE was presented by Zhou et al. (2023), who showed that synthetic data generated by oversampling underrepresented groups can mitigate algorithmic bias while keeping the predictive errors bounded.

Another option to explore new and more fair ways of algorithmic decision-making would be to use a method called SHAP instead of LIME. The SHapley Additive exPlanations (SHAP) tool is a black-box interpretation approach employed to elucidate ML predictions (Lundberg & Lee, 2017). It uses the concept of the 'Shapley value' from cooperative game theory to compute explanations of model predictions in three cases: regression values, sampling values and Quantitative Input Influence (QII) measures (Lundberg & Lee, 2017, p. 3).

To conclude, the aim of our contribution was twofold: (i) to provide a survey of algorithmic-decision systems (used mainly in criminal justice, but in other sectors too) and (ii) to discuss the importance and the crucial issues around certain ML tools that may contribute to a better understanding of algorithmic bias and effectively address issues of AI fairness promoting equity and diversity in society.

References

Ananya. (2023). Algorithms are making important decisions. What could possibly go wrong? https://www.scientificamerican.com/article/algorithms-are-making-important-decisions-what-could-possibly-go-wrong/

Andrews, D. A., & Bonta, J. (1995). *The level of service inventory - Revised.* Multi-Health Systems.

Andrews, D., Bonta, J., & Wormith, J. S. (2004). Level of service/case management inventory. PsycTESTS Dataset. https://doi.org/10.1037/t05029-000 [Data set].

Angwin, J., Larson, J., Mattu, S., & Kirchner, L. (2016). Machine bias. https://www.propublica.org/article/machine-bias-risk-assessments-in-criminal-sentencing

Arjovsky, M., Chintala, S., & Bottou, L. (2017). Wasserstein generative adversarial networks. *Proceedings of the 34th International Conference on Machine Learning* (Vol. 70, pp. 214–223). PMLR. https://proceedings.mlr.press/v70/arjovsky17a.html

Australian Human Rights Commission. (2020). Using artificial intelligence to make decisions: Addressing the problem of algorithmic bias. https://humanrights.gov.au/sites/default/files/document/publication/ahrc_technical_paper_algorithmic_bias_2020.pdf

Baeza-Yates, R. (2018). Bias on the web. *Communications of the ACM, 61*(6), 54–56. https://doi.org/10.1145/3209581

Balagopalan, A., Madras, D., Yang, D., Hadfield-Menell, D., Hadfield, G. K., & Ghassemi, M. (2023). Judging facts, judging norms: Training machine learning models to judge humans requires a modified approach to labeling data. *Science Advances, 9*(19). https://doi.org/10.1126/sciadv.abq0701

Bao, M., Zhou, A., Zottola, S. A., Brubach, B., Desmarais, S. L., Horowitz, A., & Lum, K., & Suresh Venkatasubramanian. (2021). It's COMPASlicated: The messy relationship between RAI datasets and algorithmic fairness benchmarks. arXiv: 2106.05498. https://doi.org/10.48550/arXiv.2106.05498

Barenstein, M. (2019). ProPublica's COMPAS data revisited. arXiv:1906.04711. https://doi.org/10.48550/arXiv.1906.04711

Barocas, S., & Selbst, A. D. (2016). Big Data's Disparate Impact. *California Law Review, 104*(3), 671–732. https://doi.org/10.15779/Z38BG31

Bartlett, R., Morse, A., Stanton, R., & Wallace, N. (2019). *Consumer-lending discrimination in the FinTech era.* National Bureau of Economic Research. https://doi.org/10.3386/w25943

Baumgartner, R., Arora, P., Bath, C., Burljaev, D., Ciereszko, K., Custers, B., Ding, J., Ernst, W., Fosch-Villaronga, E., Galanos, V., Gremsl, T., Hendl, T., Kropp, C., Lenk, C., Martin, P., Mbelu, S., dos Santos Bruss, S. M., Napiwodzka, K., Nowak, E., ... Williams, R. (2023). Fair and equitable AI in biomedical research and healthcare: Social science perspectives. *Artificial Intelligence in Medicine, 144*, 102658. https://doi.org/10.1016/j.artmed.2023.102658

Beaulieu, A., & Leonelli, S. (2021). *Data and society: A critical introduction.* SAGE Publications.

Bellamy, R. K. E., Mojsilovic, A., Nagar, S., Ramamurthy, K. N., Richards, J., Saha, D., Sattigeri, P., Singh, M., Varshney, K. R., Zhang, Y., Dey, K., Hind, M., Hoffman, S. C., Houde, S., Kannan, K., Lohia, P., Martino, J., & Mehta, S. (2019). AI Fairness 360: An extensible toolkit for detecting and mitigating algorithmic bias. *IBM Journal of Research and Development, 63*(4/5), 4:1–4:15. https://doi.org/10.1147/jrd.2019.2942287

Bhutta, N., Hizmo, A., & Ringo, D. (2022). How much does racial bias affect mortgage lending? Evidence from human and algorithmic credit decisions. *Board of Governors of the Federal Reserve System. Finance and Economics Discussion Series, 2022-2067*, 1–44. https://doi.org/10.17016/FEDS.2022.067

Blomberg, T., Bales, W., Mann, K., Meldrum, R., & Nedelec, J. (2010). Validation of the COMPAS risk assessment classification instrument. https://criminology.fsu.edu/sites/g/files/upcbnu3076/files/2021-03/Validation-of-the-COMPAS-Risk-Assessment-Classification-Instrument.pdf

Chakraborty, J., Majumder, S., & Menzies, T. (2021). Bias in machine learning software: Why? How? What to do? In D. Spinellis (Ed.), *Proceedings of the 29th ACM Joint Meeting on European Software Engineering Conference and Symposium on the Foundations of Software Engineering* (pp. 429–440). Association for Computing Machinery. https://doi.org/10.1145/3468264.3468537

Chawla, N. V., Bowyer, K. W., Hall, L. O., & Kegelmeyer, W. P. (2002). SMOTE: Synthetic minority over-sampling technique. *Journal of Artificial Intelligence Research, 16*, 321–357. https://doi.org/10.1613/jair.953

Chen, R. J., Wang, J. J., Williamson, D. F. K., Chen, T. Y., Lipkova, J., Lu, M. Y., Sahai, S., & Mahmood, F. (2023). Algorithmic fairness in artificial intelligence for medicine and healthcare. *Nature Biomedical Engineering, 7*, 719–742. https://doi.org/10.1038/s41551-023-01056-8.Drew.

Chouldechova, A. (2017). Fair prediction with disparate impact: A study of bias in recidivism prediction instruments. *Big Data, 5*(2), 153–163. https://doi.org/10.1089/big.2016.0047

Corbett-Davies, S., Pierson, E., Feller, A., Goel, S., & Huq, A. (2017). Algorithmic decision making and the cost of fairness. In *KDD '17: Proceedings of the 23rd ACM SIGKDD International Conference on Knowledge Discovery and Data Mining* (pp. 797–806). Association for Computing Machinery. https://doi.org/10.1145/3097983.3098095

Correa, R., Jeong, J. J., Patel, B., Trivedi, H., Gichoya, J. W., & Banerjee, I. (2023). A robust two-step adversarial debiasing with partial learning - Medical image case-studies. In B. J. Park & H. Yoshida (Eds.), *Medical Imaging 2023: Imaging Informatics for Healthcare, Research, and Applications*. https://doi.org/10.1117/12.2647285.1246908

Counts, L. (2018). Minority homebuyers face widespread statistical lending discrimination. https://newsroom.haas.berkeley.edu/minority-homebuyers-face-widespread-statistical-lending-discrimination-study-finds/

Cowgill, B., & Tucker, C. E. (2019). Economics fairness and algorithmic bias. *SSRN Electronic Journal.* https://doi.org/10.2139/ssrn.3361280

Douglas, K. S., Hart, S. D., Webster, C. D., Belfrage, H., Guy, L. S., & Wilson, C. M. (2014). Historical-Clinical-Risk Management-20, Version 3 (HCR-20V3): Development and overview. *International Journal of Forensic Mental Health, 13*(2), 93–108. https://doi.org/10.1080/14999013.2014.906519

Dressel, J., & Farid, H. (2018). The accuracy, fairness, and limits of predicting recidivism. *Science Advances, 4*(1), eaao5580. https://doi.org/10.1126/sciadv.aao5580

Emspak, J. (2016). How a machine learns prejudice. https://www.scientificamerican.com/article/how-a-machine-learns-prejudice/

Engler, A. (2019). For some employment algorithms, disability discrimination by default. https://www.brookings.edu/articles/for-some-employment-algorithms-disability-discrimination-by-default/

Engler, A. (2021). Auditing employment algorithms for discrimination. https://www.brookings.edu/articles/auditing-employment-algorithms-for-discrimination/

Fazel, S., Burghart, M., Fanshawe, T., Gil, S. D., Monahan, J., & Yu, R. (2022). The predictive performance of criminal risk assessment tools used at sentencing: Systematic review of validation studies. *Journal of Criminal Justice, 81*(101902), 101902. https://doi.org/10.1016/j.jcrimjus.2022.101902

Feldman, M., Friedler, S. A., Moeller, J., Scheidegger, C., & Venkatasubramanian, S. (2015). Certifying and removing disparate impact. In *KDD '15: Proceedings of the 21st ACM SIGKDD International Conference on Knowledge Discovery and Data Mining* (pp. 259–268). Association for Computing Machinery. https://doi.org/10.1145/2783258.2783311

48 *Rahulrajan Karthikeyan et al.*

Garrett, B. L., Jakubow, A., & Monahan, J. (2019). Judicial reliance on risk assessment in sentencing drug and property offenders: A test of the treatment resource hypothesis. *Criminal Justice and Behavior, 46*(6), 799–810. https://doi.org/10.1177/0093854819842589

Goel, S., Shroff R., Skeem J., & Slobogin, C. (2021) The accuracy, equity, and jurisprudence of criminal risk assessment. In R. Vogl (Ed.), *Research Handbook on Big Data Law* (pp. 9–28). Edward Elgar Publishing. https://doi.org/10.4337/9781788972826

Gulrajani, I., Ahmed, F., Arjovsky, M., Dumoulin, V., & Courville, A. (2017). Improved training of Wasserstein GANs. In U. von Luxburg, I. Guyon, S. Bengio, H. Wallach, & R. Fergus (Eds.), *NIPS'17: Proceedings of the 31st International Conference on Neural Information Processing Systems* (pp. 5769–5779). Curran Associates Inc. https://doi.org/10.5555/3295222.3295327

Hare, R. D. (1993). *Without Conscience: The Disturbing World of Psychopaths Among Us.* Pocket Books.

Hastie, T., Tibshirani, R., & Friedman, J. (2009). *The elements of statistical learning.* Springer. https://doi.org/10.1007/978-0-387-84858-7

Helmus, L., Thornton, D., Hanson, R. K., & Babchishin, K. M. (2011). Improving the predictive accuracy of Static-99 and Static-2002 with older sex offenders: Revised age weights. *Sexual Abuse: A Journal of Research and Treatment, 24*(1), 64–101. https://doi.org/10.1177/1079063211409951

Holm, S. (2023). Statistical evidence and algorithmic decision-making. *Synthese, 202*(1). https://doi.org/10.1007/s11229-023-04246-8

Howard, P. D. (2006). The Offender Assessment System: An evaluation of the second pilot. https://www.ojp.gov/ncjrs/virtual-library/abstracts/offender-assessment-system-evaluation-second-pilot

Huchzermeier, C., Geiger, F., Bruss, E., Godt, N., Köhler, D., Hinrichs, G., & Aldenhoff, J. B. (2007). The relationship between DSM-IV cluster B personality disorders and psychopathy according to Hare's criteria: Clarification and resolution of previous contradictions. *Behavioral Sciences & the Law, 25*(6), 901–911. https://doi.org/10.1002/bsl.722

Israni, E. T. (2017). When an Algorithm Helps Send You to Prison. https://www.nytimes.com/2017/10/26/opinion/algorithm-compas-sentencing-bias.html

Karthikeyan, R., Moore, G., Yi, C., Naveen, Y., Madhugondu, D., & Chang, Y.-Y. (2023). *Mitigating bias in judicial ML models to promote fairness in criminal justice system. Unpublished manuscript of CSE 575 course project.* Arizona State University.

Kim, P. (2017). Data-driven discrimination at work. *William and Mary Law Review, 58*(3), 857. https://scholarship.law.wm.edu/wmlr/vol58/iss3/4/

Kizilcec, R. F., & Lee, H. (2022). Algorithmic fairness in education. In W. Holmes & K. Porayska-Pomsta (Eds.), *Ethics in Artificial Intelligence in education* (pp. 174–202). Routledge.

Kleinberg, J., Ludwig, J., Mullainathan, S., & Rambachan, A. (2018). Algorithmic fairness. *AEA Papers and Proceedings.* (Vol. 108, pp. 22–27). American Economic Association. https://doi.org/10.1257/pandp.20181018

Koenecke, A., Nam, A., Lake, E., Nudell, J., Quartey, M., Mengesha, Z., Toups, C., Rickford, J. R., Jurafsky, D., & Goel, S. (2020). Racial disparities in automated speech recognition. *Proceedings of the National Academy of Sciences of the United of America, 117*(14), 7684–7689. https://doi.org/10.1073/pnas.1915768117

Lambrecht, A., & Tucker, C. (2019). Algorithmic bias? An empirical study of apparent gender-based discrimination in the display of STEM career ads. *Management Science, 65*(7), 2966–2981. https://doi.org/10.1287/mnsc.2018.3093

Langenkamp, M., Costa, A., & Cheung, C. (2019). Hiring fairly in the age of algorithms. *SSRN Electronic Journal.* https://doi.org/10.2139/ssrn.3723046

Larson, J., Mattu, S., Kirchner, L., & Angwin, J. (2016). How we analyzed the COMPAS recidivism algorithm. https://www.propublica.org/article/how-we-analyzed-the-compas-recidivism-algorithm

Latessa, E. J., Lemke, R., Makarios, M., Smith, P., & Lowenkamp, C. T. (2010). The creation and validation of the Ohio Risk Assessment System (ORAS). *Federal Probation, 74*(1), 16–22.

Latessa, E., Lovins, B., & Makarios, M. (2013). Validation of the Indiana Risk Assessment System Final Report. https://www.in.gov/courts/iocs/files/prob-risk-iras-final.pdf

Lavanchy, M. (2018). Amazon's sexist hiring algorithm could still be better than a human. https://theconversation.com/amazons-sexist-hiring-algorithm-could-still-be-better-than-a-human-105270

Lowenkamp, C. T., Johnson, J. L., Holsinger, A. M., VanBenschoten, S. W., & Robinson, C. R. (2013). The federal Post Conviction Risk Assessment (PCRA): A construction and validation study. *Psychological Services, 10*(1), 87–96. https://doi.org/10.1037/a0030343

Lundberg, S., & Lee, S. I. (2017). *A unified approach to interpreting model predictions.* arXiv:1705.07874. https://doi.org/10.48550/arXiv.1705.07874

Martinez, E., & Kirchner, L. (2021). The secret bias hidden in mortgage-approval algorithms. https://publicintegrity.org/inequality-poverty-opportunity/bias-mortgage-approval-algorithms/

Martínez, N., Vinas, A., & Matute, H. (2021). Examining potential gender bias in automated-job alerts in the Spanish market. *PLoS One, 16*(12), e0260409. https://doi.org/10.1371/journal.pone.0260409

Melton, G. B., Petrila, J., Poythress, N. G., Slobogin, C., Otto, R. K., Mossman, D., & Condie, L. O. (2018). *Psychological Evaluations for the Courts: A Handbook for Mental Health Professionals and Lawyers* (4th ed.). Guilford Press.

Park, A. (2019). Injustice Ex Machina: Predictive algorithms in criminal sentencing. https://www.uclalawreview.org/injustice-ex-machina-predictive-algorithms-in-criminal-sentencing

Pfeiffer, J., Gutschow, J., Haas, C., Möslein, F., Maspfuhl, O., Borgers, F., & Alpsancar, S. (2023). Algorithmic fairness in AI. *Business & Information Systems Engineering, 65*(2), 209–222. https://doi.org/10.1007/s12599-023-00787-x

Phenix, A., Fernandez, Y., Harris, A. J. R., Helmus, M., Hanson, R. K., & Thornton, D. (2016). Static-99R Coding Rules Revised – 2016. https://www.publicsafety.gc.ca/cnt/rsrcs/pblctns/sttc-2016/sttc-2016-en.pdf

Ribeiro, M. T., Singh, S., & Guestrin, C. (2016). Why should I trust you?: Explaining the predictions of any classifier. In *KDD '16: Proceedings of the 22nd ACM SIGKDD International Conference on Knowledge Discovery and Data Mining* (pp. 1135–1144). Association for Computing Machinery. https://doi.org/10.1145/2939672.2939778

Rieskamp, J., Hofeditz, L., Mirbabaie, M., & Stieglitz, S. (2023). Approaches to improve fairness when deploying AI-based algorithms in hiring - Using a systematic literature review to guide future research. In X. B. Tung (Ed.), *Proceedings of the 56th Hawaii International Conference on System Sciences (HICSS)*. https://scholarspace.manoa.hawaii.edu/items/606e3cb6-9ab8-44a5-9bfc-4430410bc29d

Scarlet, J. (2011). An introduction to the Psychopathy Checklist-Revised (PCL-R). https://psychopathyinfo.wordpress.com/2011/12/31/an-introduction-to-the-psychopathy-checklist-revised-pcl-r/

Suresh, H., & Guttag, J. (2021). A framework for understanding sources of harm throughout the Machine Learning life cycle. Equity and Access in Algorithms, Mechanisms, and Optimization. *Presented at the EAAMO '21: Proceedings of the 1st ACM Conference on Equity and Access in Algorithms, Mechanisms, and Optimization* (pp. 1–9). Association for Computing Machinery. https://doi.org/10.1145/3465416.3483305

Taylor, A. M. (2020). AI prediction tools claim to alleviate an overcrowded American Justice System. *But Should They be Used?* https://stanfordpolitics.org/2020/09/13/ai-prediction-tools-claim-to-alleviate-an-overcrowded-american-justice-system-but-should-they-be-used/

Vartan, S. (2019). Racial bias found in a major health care risk algorithm. https://www.scientificamerican.com/article/racial-bias-found-in-a-major-health-care-risk-algorithm/

Washington, A. L. (2019). How to argue with an algorithm: Lessons from the COMPAS-ProPublica debate. *Colorado Technology Law Journal, 17*(1), 131. http://ctlj.colorado.edu/wp-content/uploads/2021/02/17.1_4-Washington_3.18.19.pdf

World Bank Group Study. (2020). *Women, business and the law 2020*. World Bank Group.

Yang, J., Soltan, A. A. S., Eyre, D. W., Yang, Y., & Clifton, D. A. (2023). An adversarial training framework for mitigating algorithmic biases in clinical machine learning. *Npj Digital Medicine, 6*(1), 55. https://doi.org/10.1038/s41746-023-00805-y

Zhang, B. H., Lemoine, B., & Mitchell, M. (2018). Mitigating unwanted biases with adversarial learning. In *AIES '18: Proceedings of the 2018 AAAI/ACM Conference on AI, Ethics, and Society* (pp. 335–340). Association for Computing Machinery. https://doi.org/10.1145/3278721.3278779

Zhou, Y., Kantarcioglu, M., & Clifton, C. (2023). On improving fairness of AI models with synthetic minority oversampling techniques. In *Proceedings of the 2023 SIAM International Conference on Data Mining (SDM)* (pp. 874–882). https://doi.org/10.1137/1.9781611977653.ch98

Chapter 3

Ethical Challenges in the New Era of Intelligent Manufacturing Systems

Emmanouil Stathatos, Panorios Benardos and George-Christopher Vosniakos

National Technical University of Athens, Greece

Abstract

This chapter explores the ethical challenges arising from the integration of advanced artificial intelligence (AI) technologies into intelligent manufacturing systems. Machine learning (ML), augmented reality/virtual reality (AR/VR), digital twins, and human–robot collaboration (HRC) redefine industrial production, they bring forth unprecedented efficiencies and capabilities but also introduce complex ethical considerations. The text delves into issues such as data privacy, job displacement, the impact of automation on workforce dynamics, and the psychological effects of working alongside AI-powered systems. Through a detailed examination of these technologies and their implications, the chapter advocates for a dynamic ethical framework that evolves alongside technological advancements. This framework should prioritize human dignity, safety, and rights, involving all stakeholders in its development and implementation. By addressing the ethical implications of AI, AR/VR, digital twins, and HRC, the chapter underscores the necessity of balancing technological innovation with ethical responsibility. It calls for collaborative efforts involving policymakers, industry leaders, workers, and consumers to navigate the ethical landscape of intelligent manufacturing, aiming to harness the potential of these technologies responsibly for the betterment of society and the workforce.

Keywords: Artificial intelligence; machine learning; augmented/virtual reality; digital twins; human robot collaboration; ethical framework

---

The Ethics Gap in the Engineering of the Future, 51–82
Copyright © 2025 Emmanouil Stathatos, Panorios Benardos and George-Christopher Vosniakos
Published under exclusive licence by Emerald Publishing Limited
doi:10.1108/978-1-83797-635-520241004

The Advent of AI and Intelligent Manufacturing

Artificial intelligence (AI) is a scientific field that deals with the science and engineering of making intelligent machines, that is, machines that can replicate and/or emulate human intelligence. The term was first introduced by emeritus Stanford Professor John McCarthy in 1955 for the Dartmouth Summer Research Project on Artificial Intelligence (DSRPAI). However, nowadays it has become significantly more complex to define what AI is. This is due to the fact that AI encompasses theories, methods, and tools from a variety of other fields such as mathematics, computer science, and engineering. Certain enabling technologies like wireless communications, computing, and robotics are also considered to contribute to AI based on their impact regarding implementation and deployment use cases. In this context, AI applications in manufacturing started appearing approximately during the 1970s and concerned expert systems, which mimicked the decision-making process of human experts, in the form of software tools for solving order planning and logistics problems. Until the year 2000, other notable applications included artificial neural networks that act as predictive models for process and/or system performance, bio-inspired algorithms applied to optimization problems, and clustering algorithms used to determine product quality attributes. In the last 20 years, several factors have greatly contributed to redefine how AI can benefit manufacturing, resulting in accelerated adoption by the industry. Among these factors, scientific breakthroughs such as Hinton's research regarding deep belief nets, initiatives such as Industry 4.0, and technological advancements in cloud computing and sensorization can be considered as the most influential.

The integration of AI in manufacturing is not merely an advancement – it is a paradigm shift. This is echoed in research highlighting how AI, particularly machine learning (ML), is transforming the engineering workload in manufacturing processes (Csiszar et al., 2020). ML, as well as potentially AI-powered technologies such as augmented and virtual reality (AR/VR), digital twins, and human–robot collaboration (HRC), are redefining the very essence of industrial production (Bongomin et al., 2020). Unlike incremental improvements of the past, these technologies represent a transformative leap, fundamentally altering how we approach manufacturing processes (Doyle-Kent & Kopacek, 2020).

Through AI and its related technologies, we are witnessing a new era in manufacturing where efficiency, innovation, and customization are reaching unprecedented levels. This introduction paves the way for an understanding of the emerging future of manufacturing, so-called intelligent manufacturing, setting the stage for the discussion in the following sections of related ethical implications. Beside AI at its center, intelligent manufacturing typically involves further Industry 4.0 technologies, such as big data, AR, internet of things, cyber-physical systems, advanced robotics (Pereira & Romero, 2017). The integration of Industry 4.0 technologies into manufacturing processes promises significant benefits such as enhanced efficiency, reduced operational costs, and innovative product development (Schrettenbrunnner, 2020).

Important pillars of intelligent manufacturing whose ethical considerations are discussed in the next sections of this chapter are ML, AR/VR, digital twins, and HRC.

Ethical Considerations in Machine Learning in Manufacturing

At the forefront of AI revolution is ML, which enables systems to learn from data, identify patterns, and make decisions with minimal human intervention. In manufacturing, ML is pivotal in optimizing processes, reducing waste, and enhancing product quality.

Examples

In manufacturing, data that are necessary for ML may be captured from real processes through sensors working in real time in a nonintrusive monitoring mode. However, often data require intermittent measurements that do obstruct the production process, in which case drastic reduction of the amount of data, and thus interruptions, is necessary. For instance, to train a system to recognize defects in castings, a large number of photographs of defective artifacts is necessary and the type of defect needs to be recognized by an expert "teacher" to create the correct "labels" to teach to the system, see Fig. 3.1.

Such data gathering processes may be too costly to be adopted in practice, hence the transfer learning technique may be adopted, essentially "borrowing" a model pretrained (e.g., by established CNN or YOLO type artificial neural networks) with data from similar domains, for example, photos of casting defects for a different part type (Andriosopoulou et al., 2023). Thus, the availability of data that are even moderately akin to the particular problem to be solved by ML, although desirable, may raise issues of ownership as well as accountability in cases where resultant models lead to failures or pitfalls.

ML, in particular autoencoder type LSTM neural networks that deal with time series data (signals), can be used to recognize human hand movements corresponding to industrial tasks such as screwing by use of a key or a screwdriver, picking or placing parts, etc. (Mastakouris et al., 2023). Each task is made up of three or four phases, for example, in screwing a bolt: grasping the bolt, moving it

Fig. 3.1. Pressure die Casting Defects Detected by Transfer Learning Using Neural Network of CNN Type.

above the threaded hole, fitting it to the hole entrance, and rotating it by the key. Signals are generated by a smartphone attached to the wrist of the human operator and referring to accelerations in X-Y-Z axes, orientation, sound level, brightness, etc. A set of signals corresponds to a phase and it is discretized into 20 points in time. Each task with its constituent phases is repeated 125 times. LSTM networks show excellent recognition performance. They can be tied to a sliding window scanning continuous time series signals corresponding to a series of motions, thus recognizing any included motion in real time, see Fig. 3.2.

Success rate may not be sufficient for downstream exploitation, for example, by a production monitoring application or by a robot's cognition application, as this depends on the inherent variability of movements or the idiosyncratic execution of movements by different operators, but this can be remedied by capturing more signals without any particular disruption of the tasks. However, one issue that emerges is whether human workers accept such monitoring and if so, how disallowed use of such data is prevented.

Key Ethical Challenges in ML-Integrated Manufacturing

The integration of ML in manufacturing processes and systems introduces a myriad of ethical challenges that are pivotal to address. These challenges encompass not just the technological advancements but also the complex ethical landscape that comes with them.

- **Algorithmic Transparency and Bias:** ML systems in manufacturing often make critical decisions, raising the need for transparency and bias mitigation. Ensuring fairness and accountability in algorithmic decision-making is crucial

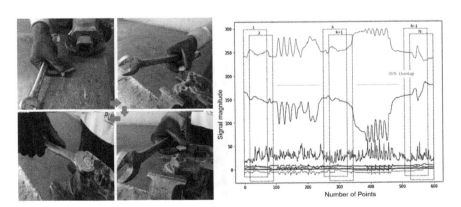

Fig. 3.2. (Left) Indicative Bolt Screwing Task Phases (Right) Typical Continuous Signal Involving Different Tasks.

to avoid discrimination and uphold ethical standards in AI applications (Hoffmann et al., 2018).

- **Data Privacy and Security:** The vast quantities of data utilized by ML systems in manufacturing bring forth significant concerns regarding privacy and data security. Protecting these data from misuse and breaches is an essential ethical obligation (Lepri et al., 2021).
- **Impact on Workforce and Society:** The implications of AI on the workforce, including job displacement, skill shifts, and psychological effects of human-AI collaboration, demand careful and ethical consideration (Gerke et al., 2020).
- **Accountability and Responsibility**: Establishing accountability in incidents involving ML systems is complex. Clear guidelines and responsibilities are vital to ethically address these concerns (Hoffmann et al., 2018).

The Imperative of Proactive Ethical Engagement

The AI industry is increasingly recognizing the importance of proactive ethical engagement, with organizations keen to address ethical issues (Stahl et al., 2022). However, there is a need for practical assistance in identifying and addressing these issues, with suggestions including embedding ethicists into development teams (McLennan et al., 2020). A research framework for implementing AI ethics in industrial settings has been proposed (Vakkuri et al., 2019), and a convergence around limited ethical principles, such as transparency, justice, and fairness, has been identified (Zhou et al., 2020). This underscores the significance of such principles in ensuring that AI in manufacturing is not only technologically sound but also ethically responsible. The EMMA (Ethical Management of AI) framework has been introduced to guide the ethical management of AI, focusing on managerial decision-making, ethical considerations, and environmental dimensions (Brendel et al., 2021).

Transparency in ML Systems

The Imperative of Transparency

In the realm of intelligent manufacturing systems, transparency in ML is not just a technical requirement but a foundational aspect of ethical operation. This is emphasized by Felzmann et al. (2020), who developed the concept of "Transparency by Design," offering practical guidance for promoting the beneficial functions of transparency in automated decision-making environments. Transparency is essential for stakeholders, ranging from operators to executives, to understand how AI systems make decisions, a key factor for trust, effective operation, and diagnosing issues. Schelenz et al. (2020) highlight that the lack of transparency can lead to user vulnerability and decreased trust, necessitating best practices for transparency in AI-based personalization systems. These insights align with the broader discussions on AI transparency in the context of intelligent manufacturing systems in Industry 4.0, as outlined by Zhou et al. (2022).

However, the effectiveness of transparency in fostering trust remains a topic of debate, as indicated by Schmidt and Biessmann (2020), who demonstrate that depending on certain personality traits, humans exhibit different susceptibilities for algorithmic bias. To enhance transparency, methods such as Explainable AI (XAI) are being explored, as suggested by Adadi and Berrada (2018), marking a significant trend in balancing the complexity of AI systems with the need for understandable and transparent operations.

Challenges to Achieving Transparency

Despite its importance, achieving transparency in AI systems is fraught with challenges. The nature of advanced ML algorithms, particularly deep learning models, often results in "black box" scenarios. These models, while powerful, lack interpretability – their decision-making processes are not easily understandable by humans. Additionally, the proprietary nature of many AI solutions adds another layer of opacity, as companies protect their algorithms as trade secrets. The following constitute some of the challenges to handle.

Complexity of Algorithms: Advanced ML algorithms, especially deep learning models, are inherently complex. Their multilayered structures and nonlinear processing make it difficult to trace how input data are transformed into outputs. This complexity is a fundamental barrier to transparency, as it challenges the ability to explain and understand AI decisions in human terms.

Proprietary Nature of AI Technologies: Many ML solutions are developed by private entities who regard their algorithms and training data as trade secrets. This proprietary approach creates a significant barrier to transparency, as the inner workings of these systems are not made available for external analysis or scrutiny.

Data Privacy and Security Concerns: Transparency in AI often requires revealing details about the data used for training and operation, which can raise privacy and security concerns. Balancing the need for transparency with the need to protect sensitive information is a complex challenge that can limit the extent to which AI systems can be made transparent.

Scalability of Transparency Solutions: As ML systems become more widespread and complex, the scalability of transparency solutions becomes a concern. Providing detailed explanations for AI decisions in large-scale systems can be resource-intensive and may not be feasible in all contexts.

Lack of Standardization: There is currently no universal standard or framework for implementing transparency in AI. This lack of standardization means that approaches to transparency can vary significantly, making it challenging to compare and assess the transparency of different AI systems.

Strategies for Enhancing Transparency

To combat these challenges, the manufacturing sector needs to adopt strategies that make AI systems more interpretable and understandable. Pertinent strategies are summarized next.

Development of Explainable AI (XAI) Techniques: This includes the development of models that are inherently interpretable, as well as tools that can explain the outputs of more complex models.

Promotion of Open-Source AI Models: Open-source models allow a broader community of developers, researchers, and users to inspect, understand, and contribute to the AI's development, fostering a more transparent environment.

Implementing Consistent Documentation and Reporting on how ML systems are designed, trained, and deployed: This includes clear documentation of data sources, model design, training processes, and performance metrics.

User-Centric Design: This involves engaging with users to understand their needs for transparency and designing interfaces and explanations that are accessible and meaningful to them.

Regulatory and Ethical Frameworks: These frameworks can set out requirements for transparency, ensuring that AI developers and users adhere to certain standards.

Collaborative Approaches: Encouraging collaboration between AI developers, users, regulatory bodies, and other stakeholders can help identify and address transparency challenges more effectively fostering the sharing of best practices and innovations.

Reliability of ML Systems

The Criticality of Reliable ML

In manufacturing, AI systems are entrusted with tasks that impact production quality, efficiency, and safety. The reliability of these systems is paramount, particularly as these systems are increasingly used in real-world often cost and safety-critical applications. An unreliable AI can lead to production errors, equipment damage, and in severe cases, pose risks to human safety. Jourdan et al. (2021) emphasize the need for continuous monitoring of ML algorithms to detect and address performance degradation over time. Ding et al. (2009) propose a control architecture that considers both reliability and safety, with a focus on adaptive optimization and safety supervision. Min et al. (2022) use recurrent disengagement events in AGVs driving tests to assess the reliability of AI systems, while Chien et al. (2020) highlight the potential of AI methods in manufacturing systems and the development of a trusted robust intelligent control strategy. Friederich and Lazarova-Molnar (2021) present a method for reliability modeling of manufacturing systems with limited data availability. Lastly, Hong et al. (Hong et al., 2021) introduce a statistical framework for AI reliability research, emphasizing the importance of test planning and data collection.

58 Emmanouil Stathatos et al.

Consequences of System Failures

When AI systems malfunction or perform suboptimally, the consequences can be extensive. Production lines may halt, leading to economic losses, or worse, they can result in workplace accidents. In high-stakes manufacturing environments, even minor errors or delays can cascade into significant operational disruptions. Consequences can be summarized as follows:

Economic Impact: The most immediate consequence of ML system failures is often economic. Production halts or slowdowns can lead to significant financial losses due to downtime and decreased productivity. In sectors with tight margins or high operational costs, these losses can be particularly severe.

Safety Hazards: ML system malfunctions in manufacturing can pose serious safety risks. This is especially true in environments where AI systems interact closely with human workers or control heavy machinery. Failures could lead to accidents, endangering worker safety and potentially leading to life-threatening situations.

Quality and Compliance Issues: System failures can result in the production of substandard products or violations of compliance standards. This can harm the company's reputation, lead to product recalls, and result in legal and regulatory consequences.

Supply Chain Disruptions: Failures in ML systems can cause disruptions that extend beyond the immediate production environment, impacting the broader supply chain. This can lead to delays in order fulfilment, affecting customer satisfaction and business relationships.

Loss of Trust: Repeated or severe failures can erode trust in ML systems among workers, management, and customers. This loss of trust can hinder the adoption of AI technologies and impede technological progress within the organization.

Measures to Enhance System Reliability

Enhancing the reliability of ML systems in manufacturing involves multiple strategies that can be summarized as follows.

Rigorous Testing and Validation: ML models should undergo extensive testing and validation before and during deployment. This includes stress testing under various conditions and ensuring that the model performs reliably across a broad spectrum of scenarios, including edge cases.

Continuous Monitoring and Maintenance: ML systems require ongoing monitoring to ensure they are functioning as intended. This involves not only tracking performance but also regularly updating and maintaining the system to adapt to new data and changing conditions.

Redundancy and Fail-Safes: Incorporating redundancy in the form of backup systems and fail-safes is crucial. This can include manual override capabilities, allowing human operators to take control if the AI system fails or behaves unpredictably.

Emergency Response Protocols: Developing clear and effective emergency response protocols can mitigate the impact of ML system failures. This involves training staff to respond quickly and efficiently to various failure scenarios.

Ethical and Legal Compliance: Ensuring that AI systems adhere to ethical guidelines and legal requirements is crucial for reliability. This includes respecting data privacy laws, adhering to industry regulations, and ensuring fairness and nondiscrimination in decisions.

Collaborative Development and Feedback Loops: Engaging with a diverse range of stakeholders, including end-users, during the development and deployment of AI systems can enhance reliability. Regular feedback loops can help identify potential issues early and ensure the system meets user needs.

Accountability for ML Failures

Navigating the Complexity of Accountability

The issue of accountability for ML failures in industry and manufacturing is a complex one, with a range of contributing factors. Nushi and Horvitz (2018) highlight the importance of understanding system behavior and the potential for flaws in conceptualization, design, and deployment. The need for methods to assess AI system performance and the importance of industry-specific accountability principles has been emphasized (Kim & Doshi-Velez, 2021; Percy et al., 2022). Research into the role of human operators and the potential for crisis communication strategies to mitigate AI failures is explored by SKITKA et al. (2000), while further investigations into this aspect are presented by Prahl and Goh (2021). McGregor (2021) proposes a practical solution by advocating for an AI Incident Database to catalog incidents and inform safety improvements. In a different approach, Angelopoulos et al. (2019) survey machine-learning solutions for fault detection, prediction, and prevention in Industry 4.0. Given this background, assigning blame when an ML system fails involves examining various (p) layers of the AI ecosystem:

- **Data Gatherers:** The quality and integrity of the data used to train ML models are critical. Data gatherers must ensure that the data are representative, unbiased, and accurately reflect the operational environment. Flaws in data can lead to flawed ML decision-making.
- **Model Designers:** ML model designers are responsible for building robust, ethical algorithms. They need to consider potential biases in their models and ensure that the ML's decision-making processes align with ethical standards.
- **Shift Managers and Operators:** Individuals who oversee ML systems in real time play a crucial role in monitoring performance and intervening when necessary. Their accountability lies in understanding the ML's capabilities and limitations and taking appropriate action when deviations occur.

- **Company Leadership:** Senior management and company leaders bear the ultimate responsibility for how ML is deployed in their operations. This includes ensuring adherence to ethical standards, compliance with regulations, and investment in proper training and oversight mechanisms.

In navigating the complexity of AI accountability, adopting a balanced approach is crucial. While recognizing that AI failures can stem from multiple sources, it is essential to consider the context of each situation. In scenarios where AI deployment is driven by profit and competitive pressures, AI companies should bear a more substantial share of the responsibility. This is especially pertinent when their products, developed in a race for market dominance, are involved in incidents. A collective responsibility model, where accountability is distributed among all stakeholders involved in the AI lifecycle, still holds value. Such a model promotes collaboration to prevent failures and ensures effective response when they occur. Nonetheless, the emphasis on AI companies' responsibility highlights the need for ethical development and deployment practices, especially in a profit-driven landscape. Establishing clear legal and policy frameworks is also vital to guide and structure the allocation of responsibility in AI-related incidents, ensuring that those most capable of controlling or mitigating risks are held accountable.

Ethical Implications of AR/VR in Industrial Production

AR and VR technologies are reshaping training and operational procedures. They provide immersive, interactive experiences that enhance understanding and improve the precision of complex manufacturing tasks.

Examples

AR can be used to demonstrate setup tasks using a digital representation of any part to be machined on any physically available machine tool in a factory (Tzimas et al., 2019). The machine is captured by a camera, digital models of fixtures being superimposed on the machine image by exploiting markers for scaling, see Fig. 3.3.

Interfacing to distance sensors imparts intelligence, while instruction scenarios are structured in terms of state threads defined in the AR program. This is a typical mix of real and digital objects (hence the augmentation) as well as limited interaction with the human operator/setter, as this is mainly a guide of manufacturing instructions. Still, it is possible to record the human's performance in terms of setup time, which opens up ethical issues of consent.

A mix of real and digital objects with much enhanced behavior of the latter according to predefined scripts is presented in (Kokkas & Vosniakos, 2019). In this case, the human plans the layout of machinery in a factory, some of the machines being physically present, while others are candidate for purchase and are being tried out in dynamic operation to see how well they fit into the existing layout, see Fig. 3.4. Operation of digital machines is again preprogrammed, but the script to be executed depends on the state of the physical system, including the

Fig. 3.3. Setting a Part (in Red) on the Table of a Real Machining Centre Using a Digital Chuck Fixture and Three Digital Clamps.

human operator's location or indeed actions, as captured by sensors or cameras. Obviously, such applications can well serve for training the workforce into new systems, hence realism in representation of objects and their behavior plays a vital role in the success of such an endeavor.

Such realism may not be necessary in AR applications where surrogate objects and actions (gestures or movements denoted as 'metaphors') are employed to avoid programming complexity or lack of detail tracking devices, for example, grasping gestures (Vasilopoulos & Vosniakos, 2021). Such simplifications may

Fig. 3.4. Layout of Real and Virtual Machinery including System State – Dependent Operations.

have a psychological or practically disruptive impact on the human immersed into the AR environment and being called to transcribe his/her experience to the real world, see Fig. 3.5. Hence the question arises whether programming the essence of the cognitive content of tasks suffices or imitating motor skills of the human operator is necessary.

Data Privacy and Security in AR/VR Systems

The Challenge of Privacy in the AR/VR Landscape

In the context of manufacturing, the use of AR/VR technologies has escalated, opening new avenues for training, design, and process management. However, these advancements bring forth significant concerns regarding data privacy. AR/VR systems can collect detailed personal information, ranging from biometric data to behavioral patterns.

The expansion of AR and VR technologies across various sectors has highlighted both their potential benefits and the challenges in adoption, particularly concerning data privacy and security. The cost implications, technical capabilities of devices, and infrastructural issues impact the adoption of these technologies, but the primary concern remains the assurance of data privacy and security for users (Kitaria & Mwadulo, 2022). Similarly, VR input devices, which enable immersive, personalized data collection, raise privacy concerns. ML-driven algorithms are proposed to countermand these privacy concerns, potentially extending to AR platforms as well (Basu et al., 2023). The ethical design and use of VR/AR products, especially in virtual learning environments, demand accountability.

Fig. 3.5. Assembly of an Impeller Into a Pump Shell by Projecting Onto the Avatar in the Virtual Environment the Tracked Human's Posture but Not His Fingers.

Designers and developers express concerns about ethical considerations in their work, emphasizing the importance of addressing these issues in educational VR/AR applications (Steele et al., 2020). Moreover, the ethical aspects of VR system design, focusing on behavioral options and the representation of reality, also require careful consideration to assess the morality of behavior within virtual environments (Brey, 1999).

Risks of Personal Data Misuse

The potential misuse of personal data in AR/VR systems presents a substantial ethical dilemma. Next, similar cases are explored, stressing the importance of user consent and strict limitations on data usage.

Inadvertent Data Collection: AR/VR systems, with their advanced sensors and tracking capabilities, can inadvertently collect more personal data than intended. This includes detailed information about users' movements, interactions, and even biometric data, which can be sensitive.

Scope Creep: There is a risk of data initially collected for benign purposes, such as training or user experience improvement, being repurposed for more invasive uses. This could include performance tracking, surveillance, or targeted advertising, often without the explicit consent of the user.

Consent and Transparency Issues: A major ethical concern is the lack of transparency and informed consent in the collection and use of personal data. Users may not be fully aware of what data are being collected, how it's being used, or who has access to it.

Discrimination and Privacy Violation: Misuse of personal data in AR/VR can lead to discrimination and violation of privacy. This includes using sensitive data to unfairly target or exclude individuals in ways that affect their employment or personal lives.

Psychological Impact: The misuse of personal data in immersive environments can have a significant psychological impact on users, including feelings of violation and mistrust, which can affect their overall well-being and attitude towards technology.

Securing AR/VR Data

Addressing the security aspect, this part will discuss the implementation of stringent data protection measures. Several strategies are crucial to prevent unauthorized data access and breaches in AR/VR integration in manufacturing as follows.

Advanced Encryption Techniques: Implementing state-of-the-art encryption is essential to protect data in transit and at rest. This includes adopting the latest standards in cryptographic security to safeguard against data interception and unauthorized access.

Robust Access Control Mechanisms: Strict access controls ensure that only authorized personnel have access to sensitive data. This includes role-based access controls, strong authentication methods, and continuous monitoring of access patterns to detect and prevent unauthorized access.

Periodic Security Audits and Vulnerability Assessments: Regular security audits and vulnerability assessments are crucial to identify and address potential security weaknesses in AR/VR systems. This includes testing for software vulnerabilities, data leakage, and intrusion attempts.

Compliance with Data Protection Regulations: Adherence to data protection laws and regulations like GDPR is critical. This involves ensuring that data collection and processing practices are compliant with legal standards, including data minimization, user consent, and rights to data access and erasure.

User Education and Awareness: Educating users about data security practices and their rights regarding personal data is important. This includes clear communication about data policies, consent mechanisms, and steps users can take to protect their own data.

Data Anonymization and Pseudonymization: Where possible, anonymizing or pseudonymizing data can reduce the risks associated with personal data misuse. This involves processing data in ways that make it difficult to associate with specific individuals.

Psychological Impacts and Gamification in AR/VR

Understanding the Psychological Effects

The immersive nature of AR/VR technologies means their use is not just a physical interaction but also a psychological one, potentially having profound sensory and cognitive effects. Users may experience a range of symptoms from mild disorientation to more intense phenomena like virtual reality sickness, which includes symptoms akin to motion sickness. Extended exposure to AR/VR environments can alter a user's perception of reality, potentially leading to a sense of detachment or confusion when transitioning back to the physical world. This disjunction can affect mental health and well-being. AR/VR can evoke strong emotional and behavioral responses, which, while beneficial in some therapeutic contexts, might be disconcerting or overwhelming in a workplace setting. Managing these emotional responses is crucial for the mental health of employees. Given these impacts, there is a need for clear guidelines on the usage of AR/VR to manage and mitigate negative psychological effects. This includes setting limits on usage duration, ensuring content quality, and providing support for users.

A range of studies have explored the psychological impacts and gamification in AR/VR in industry. Nguyen (Nguyen & Meixner, 2020) and Bucea-Manea-Ţoniş (Bucea-Manea-Ţoniş et al., 2021) and their collaborators highlight the potential of gamified AR systems in industrial settings, with the latter emphasizing the need for a safe environment. Lee (Lee et al., 2020) and Cavalcanti (Cavalcanti et al., 2021) and their teams delve into the quality of VR and its impact on user behavior, identifying content and system quality as key factors. The integration of gamification in the manufacturing industry, particularly in the context of smartphone-based gamified job design, has been shown to enhance job motivation, satisfaction, and operational performance (Liu et al., 2018). This is further supported by the development of a gamified virtual reality training environment for flexible assembly tasks, which allows for adaptive work sequences and employee-specific task fulfilment (Ulmer et al., 2020). However, the application of AR in manufacturing, while promising, is still limited due to factors such as value added, ergonomic aspects, usability, user experience, and acceptance (Zigart & Schlund, 2020).

Gamification: A Double-Edged Sword

Gamification in the workplace, particularly through AR/VR, aims to enhance productivity and employee engagement. However, it can also lead to unintended consequences. This section will explore the ethical considerations of gamification, emphasizing the need to strike a balance between motivation and employee well-being. The main points are summarized next.

Enhancing Productivity and Engagement: Gamification in AR/VR can significantly enhance productivity and engagement in the workplace. By making tasks more engaging and rewarding, it can motivate employees and improve job satisfaction.

Unintended Consequences: However, gamification can also have unintended negative consequences. For instance, prioritizing speed or other gamified metrics over safety can encourage risky behavior. There's also the risk of increased stress and burnout if employees feel constantly pressured to compete or perform.

Ethical Considerations: Implementing gamification requires careful consideration of ethical implications. This includes ensuring that gamified elements do not exploit employees, create unfair working conditions, or lead to discrimination.

Balancing Motivation and Well-being: It is crucial to strike a balance between using gamification to motivate employees and ensuring their well-being. This involves designing gamified systems that encourage positive behaviors and provide support to employees, avoiding overemphasis on competitive elements.

Inclusivity and Accessibility: Gamified AR/VR systems should be inclusive and accessible to all employees. This means considering diverse needs and preferences in the design and implementation of these systems to ensure they are beneficial and fair to everyone.

Ethical Framework for AR/VR Technologies in Industry

From an ethical standpoint, the approach to implementing AR/VR technologies must be carefully balanced. It is not just about harnessing their innovative capabilities but also about responsibly managing their implications for employees.

Human-Centric Approach and Autonomy: Emphasize designing AR/VR systems that augment human skills and foster creativity while ensuring that these technologies do not manipulate or influence users unduly. Respect for user autonomy, offering informed choices and consent options, is paramount.

Psychological Well-being and Realistic Engagement: Prioritize the psychological impacts of AR/VR, focusing on preventing negative effects. This includes realistic content portrayal, providing breaks, and avoiding overly intense experiences.

Inclusive and Diverse Design: Ensure AR/VR systems are accessible and cater to a diverse range of users. This inclusivity should extend to all aspects of design and implementation, considering various abilities, backgrounds, and preferences.

Transparent and Accountable Innovation: Maintain transparency in AR/VR operations and algorithmic functions, informing users about data usage and system functionalities. Establish clear accountability mechanisms, particularly in scenarios of data misuse or system malfunction.

Ethical Development and Ongoing Evaluation: Integrate ethical considerations continuously in the development process of AR/VR technologies. Regularly update and evaluate ethical guidelines to adapt to new challenges and advancements in the field.

Ethical Implications of Digital Twins in Manufacturing

These are advanced virtual replicas of physical systems, encompassing individual machines to entire production processes. They enable real-time monitoring, simulation, and comprehensive analysis, significantly enhancing operational efficiency.

Examples in Manufacturing

ML models may be trained with data stemming from simulations in order to replace simulation programs, as such, that may not be able to respond in real time. For example, a finite element model predicting temperatures and hence solid matter density in powder bed additive manufacturing along a laser scanning path of length 5 mm long on a single layer, see Fig. 3.6, takes 100 sec to execute, whereas an ML model with special architecture to allow parallel execution takes less than 0.1 sec (Stathatos & Vosniakos, 2019).

This makes the ML model suitable for use in a digital twin that gives in real-time information about the progress and quality of the AM process, for example, about the expected density of the part that is being built. The issue that may arise in such cases concerns consistency of inaccuracy or, equivalently, the limits of credibility of the ML model with respect to the simulation model, let alone with respect to the actual experimental data, given the usually limited possibility of extensive validation of simulation against reality. Furthermore, as digital twins typically lead to decisions made by the machine or the human, credibility of ML models becomes crucial and needs to be handled within appropriate margins of caution.

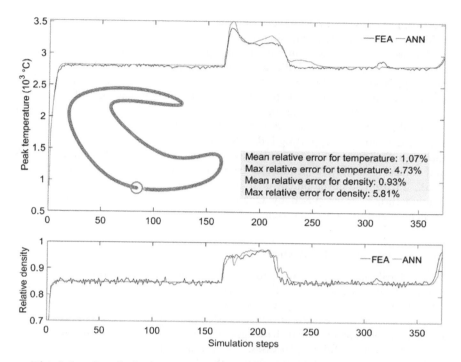

Fig. 3.6. Prediction of Temperature and Density Variation Along a Closed Laser Path in a Powder Bed Additive Manufactured Part by Simulation and by ML.

The Ethical Landscape of Digital Twins as Cyberphysical Systems

The Role of Digital Twins in Manufacturing

Digital twins in manufacturing mark a revolutionary step in how physical systems are managed and optimized. These intricate virtual replicas, embodying the essence of cyberphysical integration, have reshaped the landscape of manufacturing. They serve not just as tools for visualization but as pivotal elements in predictive maintenance, optimization of operations, and enhancement of production quality and efficiency. The integration of digital twins signifies a leap towards a more interconnected and intelligent manufacturing process, where every aspect of the physical system is mirrored, analyzed, and optimized in the digital sphere. This advancement, however, is not without its ethical considerations. The use of digital twins necessitates a thorough examination of the moral and ethical implications, particularly concerning data privacy, security, and the potential impact on employment and workforce dynamics. It's essential to explore these dimensions to understand fully the role digital twins play in modern manufacturing and the responsibilities that come with their deployment.

The use of digital twins in manufacturing presents a range of legal, methodological, and operational challenges. These include issues related to intellectual property, product liability, and data privacy (Clementson et al., 2021), as well as the need for a strong semantic foundation for accurate modeling and simulation (Park et al., 2019). Despite these challenges, digital twins offer significant potential for improving sustainability in manufacturing and maintenance (Rojek et al., 2020), and for supporting HRC (Ramasubramanian et al., 2022). However, there is a need for a more comprehensive approach that integrates human factors and considers the interdependencies between different production assets (Bécue et al., 2020). The development of a conceptual framework for digital twins in manufacturing, as proposed by Onaji et al. (2022), could help address these challenges and maximize the benefits of this technology.

Ethical Challenges in Cyberphysical Integration

The ethical landscape surrounding the deployment of digital twins in manufacturing is complex and multifaceted. As these systems act as conduits between the digital and physical worlds, they raise critical ethical questions, especially when they extend into realms of autonomy and decision-making. One of the most prominent ethical challenges arises when digital twins are employed to automate decision-making processes. This automation, while enhancing efficiency and precision, can lead to significant ethical dilemmas in scenarios involving critical safety measures and quality control. The delegation of decisions to digital twins, although beneficial in many respects, necessitates a careful examination of the ethical boundaries and responsibilities. Issues such as accountability in the event of system errors or failures, the transparency of decision-making processes, and the potential for bias in automated decisions come to the forefront. Furthermore, the extent to which these systems interact with and affect human workers raises questions about job displacement and the shifting nature of work in a digitally integrated environment. Addressing these ethical challenges requires a thoughtful approach to the design and implementation of digital twins, ensuring that they are developed and used in a manner that respects ethical standards, prioritizes human welfare, and upholds the integrity of the manufacturing process.

Digitization of Human Elements and Its Implications

The Concept of Digitizing Human Aspects

The exploration of digitizing human aspects in manufacturing goes beyond the mere creation of digital twins; it delves into the intricate interplay between technology and humanity. When a worker's physical movements, decision-making patterns, and cognitive processes are captured and replicated in a digital format, it essentially creates a virtual echo of their professional existence. This progression opens the door to numerous advancements in areas like workflow optimization, personalized training, and ergonomic assessments. For instance, by analyzing a digital twin of a worker performing a task, companies can identify potential improvements in efficiency, ergonomics, or safety. Additionally, personalized training programs can be developed using these digital models, providing employees with tailored guidance that matches their unique working style and needs.

However, alongside these benefits, the digitization of human aspects raises critical ethical concerns that must be addressed. Privacy stands at the forefront of these issues. The continuous collection of data about an individual's work habits and decisions, if not managed with strict privacy controls and consent protocols, could lead to invasive surveillance. Employees might feel their every action is being monitored and judged, leading to a stressful and distrustful work environment. There's also the risk of these data being used for purposes beyond its original intent, such as for stringent performance evaluations or even discriminatory practices.

The concept of autonomy and identity also comes into play. As digital twins replicate and perhaps even predict human behavior, there's a risk of diminishing the value of human intuition and decision-making in the manufacturing process. This could lead to a scenario where the human worker is seen more as a component of the cyberphysical system rather than as an individual with unique insights and capabilities. The challenge lies in ensuring that the digitization of human aspects in manufacturing augments and enhances human work without diminishing the worker's role and individuality.

Moreover, the blurring lines between professional and personal life in the digital realm pose another ethical challenge. Ensuring that the data collected for professional purposes do not encroach upon the personal lives and privacy of employees is paramount. Clear guidelines and transparent communication about what data are being collected, how it is being used, and who has access to it are essential to maintain trust and respect between employers and employees.

Potential Risks and Ethical Concerns

The integration of digital twins that capture human aspects in manufacturing introduces a spectrum of potential risks and ethical concerns that necessitate careful examination and management, as presented next.

Privacy and Surveillance Concerns: The continuous monitoring and digitization of an employee's work patterns present a significant risk to personal privacy. Imagine a scenario where every action, from the way an employee operates machinery to their interactions with colleagues, is tracked and recorded. These data, if used without proper consent or for purposes beyond its initial scope like performance evaluations, could lead to a sense of constant surveillance. This not only infringes on privacy but also potentially breeds a culture of mistrust and anxiety among workers, adversely affecting their morale and mental health.

Impact on Autonomy and Dignity: Digitizing human aspects in the workplace raises crucial questions about worker autonomy and dignity. When a digital twin suggests modifications to a worker's natural style or methods, it might be perceived as undermining their professional judgment and skills. This could create a workplace dynamic where human workers are seen as secondary to their digital counterparts, leading to a scenario where the value of human intuition and improvisation is downplayed. Such a shift could erode the sense of accomplishment and satisfaction that workers derive from their jobs, impacting their overall job satisfaction and self-esteem.

Misuse of Personal Data: The risk of misuse of detailed personal data is a significant ethical concern. For instance, data on an employee's decision-making patterns could be exploited for manipulative management tactics or even discriminatory practices. This misuse could manifest in several ways, such as favoritism, biased performance evaluations, or unfair workload distribution. The governance of these data, therefore, becomes paramount. Who has access to it, how it is used, and the duration of its storage are all factors that need strict regulation and transparency.

Ethical Framework for Digital Twin Technology

Addressing the concerns discussed above requires the development and implementation of robust ethical frameworks and regulatory guidelines. These should include:

Clear Consent and Transparency Protocols: Employees must be fully informed about what data are being collected, how it will be used, and who will have access to it. Their consent should be obtained in a manner that is free of coercion.

Data Governance Policies: There should be strict policies governing data access, usage, and storage, ensuring that personal data are protected and used ethically.

Worker-Centric Approach in Design: The design and implementation of these systems should prioritize the well-being and dignity of workers, ensuring that the technology augments rather than undermines their role.

Regular Ethical Audits: Regular audits and assessments should be conducted to ensure compliance with ethical standards and to identify areas for improvement.

In conclusion, while digitizing human aspects in manufacturing presents opportunities for increased efficiency and optimization, it must be approached with a strong commitment to ethical principles. This involves not only adhering to privacy and data protection laws but also considering the broader impact on worker well-being, autonomy, and workplace culture. By prioritizing ethical considerations and engaging in continuous dialogue with all stakeholders, including employees, the manufacturing industry can harness the benefits of digital twins while upholding the values of respect, dignity, and fairness in the workplace.

Ethical Dimensions of Human–Robot Collaboration in Manufacturing

The emergence of collaborative robots (cobots) is perhaps the most visible sign of intelligent manufacturing. Cobots are used alongside human workers to assemble components, combining robotic precision with human dexterity and decision-making.

Examples

In collaborative manufacturing tasks between robots and human workers, there are several interesting issues, with the main one being safety. Robots working alongside people are a potential threat as they may injure their coworkers, in extreme cases quite severely. Prescribing very conservative rules of engagement restricts freedom of collaboration and results in low productivity. Thus, machine cognition is a very significant task that allows the robot controller to make decisions, by not only recording the current situation in terms of tasks being performed but also predicting possible intended reaction steps of the human. For instance, the robot may decide to avoid being too close to the human and thus take an alternative path, once it can quickly locate the human in the common workspace, see Fig. 3.7a. Given the safety critical circumstances, reliability of sensors and software that are able to provide such information is of paramount importance (Maragkos et al., 2019). Conversely, the human needs to be warned of any dangers of collision with the robot, for example, by AR-based superposition of the robot's motion volume, see Fig. 3.7b. The human also needs to be aware of the robot's next steps, such as the rules used by the robot controller to calculate in real-time alternative paths that the robot may need to follow in order to seamlessly

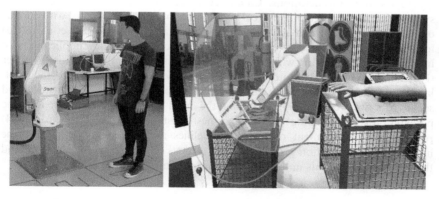

Fig. 3.7. Collaboration of Robot and Human in the Same Workspace (a) Sensor-Based Locating of the Human in Discretized Workspace (b) AR-Based Warning the Human About Next Robot Motion Volume.

72 Emmanouil Stathatos et al.

collaborate on the task (Matsas et al., 2018). Dynamic task sharing assignment or reassignment is highly desirable in many circumstances taking into account human capabilities, tiredness, mental stress, etc. AI models being able to assess or even predict such states of the human body and mind are directly related to ethics, within the application boundaries, but also outside them, sharing the same "backbone" issue of monitoring and even manipulating human behavior.

Navigating the Coexistence of Humans and AI-Enabled Coworkers

The Rise of AI-Enabled Coworkers in Manufacturing

The integration of AI-enabled robots into manufacturing signifies a ground-breaking shift in the industry, transforming the traditional landscape of automation into a realm of collaboration and interaction. In this new era, robots are no longer just tools or mechanical extensions of human labor; they are evolving into intelligent entities that work alongside humans, sharing tasks and responsibilities. This section delves into the concept of AI-enabled coworkers, highlighting the significant evolution from traditional, segregated automation to dynamic, interactive robotic systems.

The emergence of AI-enabled coworkers represents a fusion of advanced robotics, AI, and ML. These systems are designed to understand, predict, and adapt to the human work process, creating a symbiotic relationship between human and machine. The capabilities of these robots extend beyond repetitive tasks to include decision-making, learning from experience, and even collaborating on complex tasks. This evolution necessitates a rethinking of the manufacturing floor – not just in terms of technological infrastructure but also in the context of workflow, safety protocols, and workforce training.

The ethical aspects of HRC are a complex and evolving field, with a range of typologies and considerations. Callari et al. (2023) emphasize the need for a systemic approach to guiding ethical behavior in human-robot teams, with a focus on role establishment and accountability. Sequeira (2019) and Riek et al. (2015) highlight the importance of considering the societal and policy implications of human–robot interaction, with a call for cross-disciplinary collaboration. Fletcher (Fletcher & Webb, 2017) and Ostrowski (Ostrowski et al., 2022) and their collaborators discuss the specific challenges and considerations in industrial and social robotics, respectively, including the need for equitable design and the potential for perpetuating social inequities. Lastly, Riek and Howard (2014) propose the development of ethical guidelines and a code of ethics for the human–robot interaction profession, with a focus on safety, risk evaluation, and the affordance of rights and protections in human-robot interactions.

Transition to Collaborative Work Environments

The advent of AI-enabled coworkers ushers in a transformative change in the manufacturing environment. This transition brings a new dimension to the nature of work, fundamentally altering how human workers interact with their

Ethical Challenges in the New Era 73

mechanical counterparts. In these collaborative work environments, the skill sets required from human workers evolve. There is a growing need for skills such as programming, system troubleshooting, and data analysis, alongside traditional manufacturing skills.

The collaboration with AI-enabled robots also demands a shift in the work culture and mindset. Human workers need to adapt to sharing their workspace with machines that can learn, make decisions, and, in some cases, operate autonomously. This requires a new understanding of teamwork, where collaboration extends beyond human-to-human interaction to include human-to-machine cooperation. The psychological and social aspects of this shift are as significant as the technical ones. Workers must not only trust in the reliability and safety of these systems but also adjust to the changing nature of their roles and responsibilities.

Furthermore, the transition to collaborative work environments raises questions about job security and the potential displacement of traditional roles. While AI-enabled coworkers can increase efficiency and productivity, there is a concern about the impact on employment opportunities. Addressing these concerns involves not only technological innovation but also proactive workforce development and policy interventions.

In summary, the rise of AI-enabled coworkers in manufacturing marks a pivotal moment in the industry's evolution. It presents opportunities for increased efficiency, innovation, and reshaping of work roles. However, it also brings challenges that need careful consideration and management, including workforce training, adaptation to new work cultures, and addressing concerns about employment impacts. This transition to collaborative work environments is not just a technological leap but a step into a new era of manufacturing where the lines between human and machine capabilities are increasingly blurred.

Safeguarding Physical and Psychological Health in HRC

Ensuring Personnel Safety

In the realm of HRC, the physical safety of human workers stands as a paramount concern. The close interaction between humans and robots in shared workspaces introduces new dimensions of risk, necessitating comprehensive safety measures. This section explores the various strategies and technologies essential for ensuring personnel safety in HRC environments.

Key to this safety paradigm are advanced sensing technologies and safety protocols integrated within robotic systems. These technologies include collision detection sensors, force limiters, and emergency stop mechanisms that allow robots to operate safely alongside humans. They are designed to promptly respond to potential hazards, thereby reducing the risk of accidents. Additionally, the layout of the workspace plays a crucial role, with designs that facilitate safe human–robot interaction and minimize risk-prone scenarios.

Beyond technological solutions, personnel safety in HRC settings also hinges on thorough training and education of workers. This training should encompass

74 Emmanouil Stathatos et al.

not only the operational aspects of working with AI-enabled robots but also safety protocols and emergency response procedures. Workers need to be adept at recognizing potential hazards and understanding the limitations and capabilities of their robotic coworkers to foster a safe working environment.

Addressing Psychological Impacts

While ensuring physical safety is critical, the psychological well-being of workers in HRC settings is equally important. The introduction of AI-enabled coworkers into the manufacturing space can profoundly impact the psychological state of human workers. This subsection examines these impacts, focusing on worker morale, job satisfaction, and the sense of agency.

Working alongside intelligent machines can elicit feelings of insecurity or anxiety among some workers. There might be concerns about job displacement, or feelings of being overshadowed by the capabilities of AI-enabled robots. Moreover, the shift in work dynamics can lead to a perceived loss of control or agency in decision-making processes, impacting job satisfaction and self-efficacy.

To address these psychological impacts, it's essential to foster a work environment that values and emphasizes the unique contributions of human workers. This involves clear communication about the role of AI-enabled coworkers, reassurance about job security, and highlighting the collaborative nature of these new work arrangements. Additionally, providing opportunities for workers to upskill and adapt to the evolving work environment can help mitigate feelings of redundancy and enhance job satisfaction.

Moreover, psychological support mechanisms, such as counselling services or stress management programs, can be beneficial. These initiatives can assist workers in navigating the changes and challenges brought about by HRC, ensuring their mental and emotional well-being is maintained.

Socio-Economic Implications of HRC

Potential Job Displacement and Inequality

The incorporation of AI-enabled robots into manufacturing processes represents a significant shift in the industrial landscape, bringing with it complex socio-economic implications. One of the most pressing concerns is the potential for job displacement. As robots become capable of performing tasks that were previously resistant to automation, the nature of employment in manufacturing is poised to undergo profound changes. This evolution is not just about the replacement of manual labor with machines; it extends to skilled tasks that require decision-making and problem-solving abilities, traditionally considered the domain of human workers.

This shift raises critical questions about the future of employment in the manufacturing sector. There's a potential for significant job displacement, especially for roles that robots can perform more efficiently. While new job opportunities in robot maintenance, programming, and supervision may arise, these

Ethical Challenges in the New Era 75

roles often require skills that the current workforce might not possess. This mismatch can lead to a widening of socio-economic inequalities, as those unable to adapt or retrain could find themselves marginalized in the job market.

The economic impacts of HRC extend beyond individual job loss to broader societal implications. Regions heavily reliant on manufacturing may face economic challenges, and there could be a ripple effect on other sectors. Policymakers and industry leaders must consider these factors, working towards solutions that balance technological advancement with economic stability and workforce development.

Dehumanization Concerns in the Workplace

Alongside the economic considerations, there are significant concerns about the dehumanization of the workforce in HRC settings. As robots take on more complex and varied tasks, there's a risk that human workers may be perceived as less essential or valuable. This perception could lead to a workplace environment where humans are expected to match the efficiency and consistency of machines, disregarding the inherent differences between human and robotic workers.

This push towards a machine-like work ethic not only overlooks the unique strengths and capabilities of human workers but also risks creating a dehumanized work environment. In such settings, the emphasis on efficiency and productivity could overshadow the importance of creativity, problem-solving, and human judgment. It's essential to recognize and value the distinct contributions that human workers bring to the collaborative process, ensuring that the workplace remains a space where human dignity, creativity, and well-being are prioritized.

To address these concerns, it's important to foster a workplace culture that respects and integrates the human element within HRC. This involves creating policies and practices that recognize the value of human workers, ensuring that their roles evolve alongside technological advancements rather than being overshadowed by them. Additionally, there should be a focus on developing human-centric design principles in robotic systems, emphasizing collaboration and support rather than replacement.

In conclusion, the socio-economic implications of HRC in manufacturing are multifaceted and far-reaching. The challenges of potential job displacement, inequality, and dehumanization in the workplace require thoughtful consideration and proactive management. By addressing these issues holistically, it's possible to harness the benefits of HRC while maintaining a balanced, equitable, and human-centered work environment.

Ethical Framework for Human–Robot Collaboration

In conclusion, establishing an ethical framework for HRC in manufacturing is imperative to navigate the complex interplay of technology, society, and human values. This framework should encapsulate principles that prioritize human

76 Emmanouil Stathatos et al.

dignity, safety, and well-being while harnessing the benefits of AI and robotics. Key components of this framework include:

Human-Centric Design: Ensuring that AI-enabled robots are designed with the primary goal of augmenting human capabilities, not replacing them. This involves integrating ergonomic considerations, intuitive interfaces, and systems that complement human skills and creativity.

Safety and Reliability: Emphasizing rigorous safety standards and reliable performance of robotic systems to protect human workers from physical harm. This encompasses advanced sensing technologies, emergency protocols, and continuous monitoring for potential hazards.

Transparency and Accountability: Maintaining transparent operations and clear accountability in HRC scenarios. This means having well-defined roles, responsibilities, and decision-making processes where the actions of both humans and robots are understandable and accountable.

Ethical Training and Education: Providing comprehensive training for workers not only in operational procedures but also in understanding the ethical implications of working with AI-enabled robots. This education should foster awareness of potential biases, the importance of equitable treatment, and respect for diversity in the workplace.

Psychological Support and Job Security: Addressing the psychological impacts of HRC by offering support mechanisms for workers adapting to new roles and technologies. This includes safeguarding job security through upskilling programs and policies that mitigate the risk of job displacement and economic inequality.

Inclusive Policymaking: Involving a diverse range of stakeholders, including workers, ethicists, technologists, and policymakers, in the development of guidelines and regulations for HRC. This ensures that diverse perspectives are considered, leading to more equitable and effective policies.

Socio-Economic Balance: Striving for a balance between technological advancement and socio-economic stability. This means addressing potential job displacement and economic disparities while exploring new opportunities for employment and growth in the era of AI and robotics.

Respect for Human Autonomy and Agency: Ensuring that AI-enabled robots support rather than undermine human decision-making and autonomy. This involves creating systems that enhance human judgment and problem-solving rather than dictating or overriding human choices.

Continuous Ethical Evaluation: Regularly reviewing and updating the ethical framework to adapt to emerging challenges and advancements in AI and robotics. This dynamic approach allows for the integration of new insights and addresses evolving societal needs and technological capabilities.

Epilogue: Charting an Ethical Course in Intelligent Manufacturing

As we conclude this exploration into the ethical challenges of intelligent manufacturing, it is evident that the integration of advanced AI technologies is transforming the manufacturing landscape. This transformation, while promising unparalleled efficiencies and capabilities, also brings to light a spectrum of ethical considerations that demand our attention and action.

The discussions in this chapter underscore the necessity of a steadfast commitment to ethical principles in the face of rapid technological advancement. It is imperative that the pursuit of innovation is not at the expense of human dignity, safety, and rights. As we have seen in the various sections, each technological advancement, be it AI systems, AR/VR, digital twins, or HRC, carries its own set of ethical implications that need to be carefully navigated.

The evolving nature of intelligent manufacturing systems requires an equally dynamic ethical framework. Such a framework should be proactive, evolving concurrently with technological advancements. It should encompass principles of fairness, accountability, transparency, and respect for human rights. This framework must be inclusive, involving all stakeholders – from policymakers and industry leaders to workers and consumers – in the dialogue and decision-making processes.

The development and implementation of ethical guidelines in intelligent manufacturing should not be the sole responsibility of any single entity. Instead, it calls for a collaborative effort involving various stakeholders. Governments should provide a framework that encourages ethical innovation, while avoiding overly prescriptive regulations that could stifle progress or diminish individual rights. Industry leaders are expected to self-regulate, integrating ethical considerations into their corporate strategies as a competitive advantage. Workers should proactively engage in dialogue to voice their perspectives and safeguard their interests. Similarly, consumers have the power to influence industry practices by choosing products and services that align with ethical standards, driving change through market demand.

As we venture further into this new era of intelligent manufacturing, we carry with us both optimism and caution. Optimism for the potential these technologies hold to revolutionize manufacturing and improve lives, and caution to ensure that these advancements are harnessed responsibly. By placing ethical considerations at the forefront of technological innovation, we can strive towards a future where technology enhances human work, creativity, and well-being, creating a harmonious balance between efficiency and ethics.

References

Adadi, A., & Berrada, M. (2018). Peeking Inside the Black-Box: A Survey on Explainable Artificial Intelligence (XAI). *IEEE Access, 6*, 52138–52160. https://doi.org/10.1109/ACCESS.2018.2870052

Andriosopoulou, G., Mastakouris, A., Masouros, D., Benardos, P., Vosniakos, G. C., & Soudris, D. (2023). Defect recognition in high-pressure die-casting parts using neural networks and transfer learning. *Metals, 13*, 1104. https://doi.org/10.3390/MET13061104

Angelopoulos, A., Michailidis, E. T., Nomikos, N., Trakadas, P., Hatziefremidis, A., Voliotis, S., & Zahariadis, T. (2019). Tackling faults in the industry 4.0 era—A survey of machine-learning solutions and key aspects. *Sensors, 20*(1), 109. https://doi.org/10.3390/s20010109

Basu, A., Sunny, M. J. M., & Guthula, J. S. N. (2023). *Privacy concerns from variances in spatial navigability in VR.* https://arxiv.org/abs/2302.02525

Bécue, A., Maia, E., Feeken, L., Borchers, P., & Praça, I. (2020). A new concept of digital twin supporting optimization and resilience of factories of the future. *Applied Sciences, 10*(13), 4482. https://doi.org/10.3390/app10134482

Bongomin, O., Gilibrays Ocen, G., Oyondi Nganyi, E., Musinguzi, A., & Omara, T. (2020). Exponential disruptive technologies and the required skills of industry 4.0. *Journal of Engineering, 2020*, 1–17. https://doi.org/10.1155/2020/4280156

Brendel, A. B., Mirbabaie, M., Lembcke, T.-B., & Hofeditz, L. (2021). Ethical Management of Artificial Intelligence. *Sustainability, 13*(4), 1974. https://doi.org/10.3390/su13041974

Brey, P. (1999). The ethics of representation and action in virtual reality. *Ethics and Information Technology, 1*(1), 5–14. https://doi.org/10.1023/A:1010069907461

Bucea-Manea-Ţoniş, R., Gurgu, E., Martins, O. M. D., & Simion, V. E. (2021). An overview of how VR/AR applications assist specialists in developing better consumer behavior and can revolutionize our life. In *Consumer Happiness: Multiple Perspectives. Studies in Rhythm Engineering* (pp. 231–253). Springer. https://doi.org/10.1007/978-981-33-6374-8_12

Callari, T., Oostveen, A.-M., Hubbard, E.-M., Fletcher, S., & Lohse, N. (2023). Where are we at? A review of the advances in the ethical aspects of human-robot collaboration. In *14th International Conference on. Applied Human Factors and Ergonomics (AHFE 2023)*. https://doi.org/10.54941/ahfe1003506

Cavalcanti, J., Valls, V., Contero, M., & Fonseca, D. (2021). Gamification and hazard communication in virtual reality: A qualitative study. *Sensors, 21*(14), 4663. https://doi.org/10.3390/s21144663R

Chien, C.-F., Dauzère-Pérès, S., Huh, W. T., Jang, Y. J., & Morrison, J. R. (2020). Artificial intelligence in manufacturing and logistics systems: Algorithms, applications, and case studies. *International Journal of Production Research, 58*(9), 2730–2731. https://doi.org/10.1080/00207543.2020.1752488

Clementson, J., Teng, J., Wood, P., & Windmill, C. (2021). Legal considerations for using digital twins in additive manufacture – A review of the literature. In *Volume 15: Advances in Manufacturing Technology (XXXIV*, pp. 91–96). https://doi.org/10.3233/ATDE210018

Csiszar, A., Hein, P., Wachter, M., Verl, A., & Bullinger, A. C. (2020). Towards a user-centered development process of machine learning applications for manufacturing domain experts. In *2020 Third International Conference on Artificial Intelligence for Industries (AI4I)* (pp. 36–39). IEEE. https://doi.org/10.1109/AI4I49448.2020.00015

Ding, H., Kain, S., Schiller, F., & Stursberg, O. (2009). Increasing reliability of intelligent manufacturing systems by adaptive optimization and safety supervision. *IFAC Proceedings Volumes, 42*(8), 1533–1538. https://doi.org/10.3182/20090630-4-ES-2003.00250

Doyle-Kent, M., & Kopacek, P. (2020). Industry 5.0: Is the manufacturing industry on the cusp of a new revolution? In *Lecture Notes in Mechanical Engineering* (pp. 432–441). https://doi.org/10.1007/978-3-030-31343-2_38

Felzmann, H., Fosch-Villaronga, E., Lutz, C., & Tamò-Larrieux, A. (2020). Towards transparency by design for artificial intelligence. *Science and Engineering Ethics*, *26*(6), 3333–3361. https://doi.org/10.1007/s11948-020-00276-4

Fletcher, S. R., & Webb, P. (2017). Industrial robot ethics: The challenges of closer human collaboration in future manufacturing systems. In *A World with robots* (pp. 159–169). https://doi.org/10.1007/978-3-319-46667-5_12

Friederich, J., & Lazarova-Molnar, S. (2021). Process mining for reliability modeling of manufacturing systems with limited data availability. In *2021 8th International Conference on Internet of Things: Systems, Management and Security (IOTSMS)* (pp. 1–7). IEEE. https://doi.org/10.1109/IOTSMS53705.2021.9704921

Gerke, S., Minssen, T., & Cohen, G. (2020). Ethical and legal challenges of artificial intelligence-driven healthcare. In *Artificial Intelligence in healthcare*. INC. https://doi.org/10.1016/B978-0-12-818438-7.00012-5

Hoffmann, A. L., Roberts, S. T., Wolf, C. T., & Wood, S. (2018). Beyond fairness, accountability, and transparency in the ethics of algorithms: Contributions and perspectives from LIS. *Proceedings of the Association for Information Science and Technology*, *55*(1), 694–696. https://doi.org/10.1002/pra2.2018.14505501084

Hong, Y., Lian, J., Xu, L., Min, J., Wang, Y., Freeman, L. J., & Deng, X. (2021). *Statistical perspectives on reliability of Artificial Intelligence systems*. https://arxiv.org/abs/2111.05391

Jourdan, N., Sen, S., Husom, E. J., Garcia-Ceja, E., Biegel, T., & Metternich, J. (2021). *On the reliability of machine learning applications in manufacturing environments*. https://arxiv.org/abs/2112.06986

Kim, B., & Doshi-Velez, F. (2021). Machine learning techniques for accountability. *AI Magazine*, *42*(1), 47–52. https://doi.org/10.1002/j.2371-9621.2021.tb00010.x

Kitaria, D., & Mwadulo, M. (2022). Adoption of Augmented Reality (AR) and Virtual Reality (VR) in healthcare systems. *African Journal of Science, Technology and Social Sciences*, *1*(1). https://doi.org/10.58506/ajstss.v1i1.70

Kokkas, A., & Vosniakos, G.-C. (2019). An Augmented Reality approach to factory layout design embedding operation simulation. *International Journal on Interactive Design and Manufacturing*, *13*, 1061–1071.

Lee, M., Lee, S. A., Jeong, M., & Oh, H. (2020). Quality of virtual reality and its impacts on behavioral intention. *International Journal of Hospitality Management*, *90*, 102595. https://doi.org/10.1016/j.ijhm.2020.102595

Lepri, B., Oliver, N., & Pentland, A. (2021). Ethical machines: The human-centric use of artificial intelligence. *iScience*, *24*(3), 102249. https://doi.org/10.1016/j.isci.2021.102249

Liu, M., Huang, Y., & Zhang, D. (2018). Gamification's impact on manufacturing: Enhancing job motivation, satisfaction and operational performance with smartphone-based gamified job design. *Human Factors and Ergonomics in Manufacturing & Service Industries*, *28*(1), 38–51. https://doi.org/10.1002/hfm.20723

Maragkos, C., Vosniakos, G.-C., & Matsas, E. (2019). Virtual reality assisted robot programming for human collaboration. *Procedia Manufacturing*, *38*, 1697–1704.

Mastakouris, A., Andriosopoulou, G., Masouros, D., Benardos, P., Vosniakos, G.-C., & Soudris, D. (2023). Human worker activity recognition in a production floor environment through deep learning. *Journal of Manufacturing Systems*, *71*, 115–130.

Matsas, E., Vosniakos, G.-C., & Batras, D. (2018). Prototyping proactive and adaptive techniques for human-robot collaboration in manufacturing using virtual reality. *Robotics and Computer-Integrated Manufacturing, 50*, 168–180.

McGregor, S. (2021). Preventing repeated real world AI failures by cataloging incidents: The AI incident database. *Proceedings of the AAAI Conference on Artificial Intelligence, 35*(17), 15458–15463. https://doi.org/10.1609/aaai.v35i17.17817

McLennan, S., Fiske, A., Celi, L. A., Müller, R., Harder, J., Ritt, K., Haddadin, S., & Buyx, A. (2020). An embedded ethics approach for AI development. *Nature Machine Intelligence, 2*(9), 488–490. https://doi.org/10.1038/s42256-020-0214-1

Min, J., Hong, Y., King, C. B., & Meeker, W. Q. (2022). Reliability analysis of artificial intelligence systems using recurrent events data from autonomous vehicles. *Journal of the Royal Statistical Society - Series C: Applied Statistics, 71*(4), 987–1013. https://doi.org/10.1111/rssc.12564

Nguyen, D., & Meixner, G. (2020). A survey of gamified augmented reality systems for procedural tasks in industrial settings. *IFAC-PapersOnLine, 53*(2), 10096–10100. https://doi.org/10.1016/j.ifacol.2020.12.2733

Nushi, B., Kamar, E., & Horvitz, E. (2018). Towards accountable AI: Hybrid human-machine analyses for characterizing system failure. *Proceedings of the AAAI Conference on Human Computation and Crowdsourcing, 6*, 126–135. https://doi.org/10.1609/hcomp.v6i1.13337

Onaji, I., Tiwari, D., Soulatiantork, P., Song, B., & Tiwari, A. (2022). Digital twin in manufacturing: Conceptual framework and case studies. *International Journal of Computer Integrated Manufacturing, 35*(8), 831–858. https://doi.org/10.1080/0951192X.2022.2027014

Ostrowski, A. K., Walker, R., Das, M., Yang, M., Breazea, C., Park, H. W., & Verma, A. (2022). Ethics, equity, & justice in human-robot interaction: A review and future directions. In *2022 31st IEEE International Conference on Robot and Human Interactive Communication (RO-MAN)* (pp. 969–976). IEEE. https://doi.org/10.1109/RO-MAN53752.2022.9900805

Park, H., Easwaran, A., & Andalam, S. (2019). Challenges in digital twin development for cyber-physical production systems. In *Lecture Notes in Computer Science (including subseries Lecture Notes in Artificial Intelligence and Lecture Notes in Bioinformatics)* (pp. 28–48). https://doi.org/10.1007/978-3-030-23703-5_2

Percy, C., Dragicevic, S., Sarkar, S., & d'Avila Garcez, A. (2022). Accountability in AI: From principles to industry-specific accreditation. *AI Communications, 34*(3), 181–196. https://doi.org/10.3233/AIC-210080

Pereira, A. C., & Romero, F. (2017). A review of the meanings and the implications of the Industry 4.0 concept. *Procedia Manufacturing, 13*, 1206–1214. https://doi.org/10.1016/j.promfg.2017.09.032

Prahl, A., & Goh, W. W. P. (2021). "Rogue machines" and crisis communication: When AI fails, how do companies publicly respond? *Public Relations Review, 47*(4), 102077. https://doi.org/10.1016/j.pubrev.2021.102077

Ramasubramanian, A. K., Mathew, R., Kelly, M., Hargaden, V., & Papakostas, N. (2022). Digital twin for human–robot collaboration in manufacturing: Review and outlook. *Applied Sciences, 12*(10), 4811. https://doi.org/10.3390/app12104811

Riek, L. D., Hartzog, W., Howard, D. A., Moon, Aj, & Calo, R. (2015). The emerging policy and ethics of human robot interaction. In *Proceedings of the Tenth*

Annual ACM/IEEE International Conference on Human-Robot Interaction Extended Abstracts (pp. 247–248). ACM. https://doi.org/10.1145/2701973.2714393

Riek, L. D., & Howard, D. (2014). A code of ethics for the human-robot interaction profession. *We Robot Conference*, 1–10.

Rojek, I., Mikołajewski, D., & Dostatni, E. (2020). Digital twins in product lifecycle for sustainability in manufacturing and maintenance. *Applied Sciences, 11*(1), 31. https://doi.org/10.3390/app11010031

Schelenz, L., Segal, A., & Gal, K. (2020). *Applying transparency in artificial intelligence based personalization systems.* https://arxiv.org/abs/2004.00935

Schmidt, P., & Biessmann, F. (2020). *Calibrating human-AI collaboration: Impact of risk, ambiguity and transparency on algorithmic bias* (pp. 431–449). https://doi.org/10.1007/978-3-030-57321-8_24

Schrettenbrunnner, M. B. (2020). Artificial-intelligence-driven management. *IEEE Engineering Management Review, 48*(2), 15–19. https://doi.org/10.1109/EMR.2020.2990933

Sequeira, J. S. (2019). Ethics in human-robot interaction. In *Emotional design in human-robot interaction* (pp. 111–116). https://doi.org/10.1007/978-3-319-96722-6_7

Skitka, L. J., Mosier, K., & Burdick, M. D. (2000). Accountability and automation bias. *International Journal of Human-Computer Studies, 52*(4), 701–717. https://doi.org/10.1006/ijhc.1999.0349

Stahl, B. C., Antoniou, J., Ryan, M., Macnish, K., & Jiya, T. (2022). Organisational responses to the ethical issues of artificial intelligence. *AI & Society, 37*(1), 23–37. https://doi.org/10.1007/s00146-021-01148-6

Stathatos, E., & Vosniakos, G.-C. (2019). Real-time simulation for long paths in laser-based additive manufacturing: A machine learning approach. *The International Journal of Advanced Manufacturing Technology, 104*(5–8), 1967–1984. https://doi.org/10.1007/s00170-019-04004-6

Steele, P., Burleigh, C., Kroposki, M., Magabo, M., & Bailey, L. (2020). Ethical considerations in designing virtual and augmented reality products—Virtual and augmented reality design with students in mind: designers' perceptions. *Journal of Educational Technology Systems, 49*(2), 219–238. https://doi.org/10.1177/0047239520933858

Tzimas, E., Vosniakos, G.-C., & Matsas, E. (2019). Machine tool setup instructions in the smart factory using augmented reality: A system construction perspective. *International Journal on Interactive Design and Manufacturing, 13*, 121–136.

Ulmer, J., Braun, S., Cheng, C.-T., Dowey, S., & Wollert, J. (2020). Gamified virtual reality training environment for the manufacturing industry. In *2020 19th International Conference on Mechatronics - Mechatronika (ME)* (pp. 1–6). IEEE. https://doi.org/10.1109/ME49197.2020.9286661

Vakkuri, V., Kemell, K.-K., & Abrahamsson, P. (2019). *AI ethics in industry: A research framework.* https://arxiv.org/abs/1910.12695

Vasilopoulos, G., & Vosniakos, G.-C. (2021). Preliminary design of assembly system and operations for large mechanical products using a game engine. *Procedia CIRP, 104*, 1395–1400.

Zhou, J., Chen, F., Berry, A., Reed, M., Zhang, S., & Savage, S. (2020). A survey on ethical principles of AI and implementations. In *2020 IEEE Symposium Series on Computational Intelligence (SSCI)* (pp. 3010–3017). IEEE. https://doi.org/10.1109/SSCI47803.2020.9308437

Zhou, L., Jiang, Z., Geng, N., Niu, Y., Cui, F., Liu, K., & Qi, N. (2022). Production and operations management for intelligent manufacturing: A systematic literature review. *International Journal of Production Research*, *60*(2), 808–846. https://doi.org/10.1080/00207543.2021.2017055

Zigart, T., & Schlund, S. (2020). Evaluation of augmented reality technologies in manufacturing – A literature review. *Advances in Intelligent Systems and Computing Book Series (AISC)*, *1207*, 75–82. https://doi.org/10.1007/978-3-030-51369-6_11

Chapter 4

From Croesus to Computers: Logic of Perverse Instantiation

Goran Rujević

University of Novi Sad, Serbia

Abstract

Perverse instantiation is one of many hypothetical failure modes of AI, specifically one in which the AI fulfils the command given to it by its principal in a way which is both unforeseen and harmful. A lot is already said about perverse instantiation itself, especially when such a failure mode presents an existential risk, as would be the case with a superintelligent AI. However novel these disaster scenarios may be, similar fictional cautionary tales already exist in many cultures: tragic stories about misinterpreted prophecies and grand wishes gone awry, from Croesus to Macbeth. Analysis of both old and new tales of perverse instantiation reveals that the core of the issue is an ancient philosophical and logical problem that even Socrates faced: the problem of defining terms. Unlike the Socratic problem, which focused on finding a good intensional definition, perverse instantiation encompasses problems that arise from both badly defined intension of terms (their internal content) and badly defined extension of terms (their range of applicability). However, models of machine learning that use vast amounts of training data hold the promise of resolving the issue of badly defined extension of terms. The issue of defining intension of terms remains. Further parallels can be found between scenarios of perverse instantiation and Socrates' dialogues with obstinate sophists, such as importance of philosophical reflection and discussion. This indicates that our future challenges in working with AI may still have a lot to do with retracing Socrates' steps.

Keywords: AI ethics; extension; intension; perverse instantiation; Socrates; sophists

---

The Ethics Gap in the Engineering of the Future, 83–104
Copyright © 2025 Goran Rujević
Published under exclusive licence by Emerald Publishing Limited
doi:10.1108/978-1-83797-635-520241005

84 Goran Rujević

Every year for the past 17 years, The Association for Computational Heresy has held the SIGBOVIK conference, a humorous event spoofing academic conferences in the field of computer science with all the bells and whistles from paper presentations to official proceedings. The comedic spirit of the SIGBOVIK is plainly evident from proclamations of covering such arcane topics as natural intelligence, artificial stupidity, thaughmaturgic [*sic*] circle, as well as from the fact that it is taking place on April 1st of every year. However, at the SIGOBIVK 2013, Tom Murphy submitted a paper that was indeed entertaining, but also contained a disclaimer that 'This work is 100% real'. (Murphy, 2013, p. 112)

The paper *The First Level of Super Mario Bros. is Easy with Lexicographic Orderings and Time Travel... after that it gets a little tricky* presents Murphy's experience with trying to find an approach to automate the playing of multiple games of the Nintendo Entertainment System. The process involved a sample of human-played game and an algorithm that would attempt to extract a winning function from the recorded memory states, which would then be applied to automated play attempts. It should be noted that the playing machine was not limited only to in-game moves, but had access to every possible input a human user could make. Some machine-playing attempts were utter dead ends, some stumbled upon and exploited unusual glitches in the games, but by far the most interesting result was encountered with the game of *Tetris*.

As one might expect, the machine was abysmal at the game. *Tetris* is one of those games that do not have so much of a winning state but rather revolve around constant striving to avoid losing, that is, encountering the 'game over' screen. As mistakes literally pile up on top of each other, the best one can hope for is a high score at the moment of defeat. The playing machine was, of course, blind to such specificities, it could only recognize moves that incremented the total score. Since every placed block did increase the score somewhat, that is what the machine did, which lead to a haphazardly stacked tower quickly reaching the top of the screen. Moments before the losing block would land, an unforeseen event happened – the machine paused the game indefinitely. In essence, this unconventional move did prevent the 'game over' screen from ever appearing. But should this approach be accepted as a winning strategy? Murphy commented on this that in *Tetris* 'Truly, the only winning move is not to play' (Murphy, 2013, p. 131).

On the one hand, we could commend Murphy's *playfun* algorithm for 'creative problem-solving'; on the other hand, we could scold it for twisting both the letter and the spirit of the notion of 'winning'. Either way, what happened with the automated *Tetris* game was certainly a result that very few, if any, could expect beforehand. One could even say that this was a case of misalignment between what we expected an automated system might do and what the system actually did. In this case, the subverted expectation was practically inconsequential and indeed amusing and most apt for a satirical conference. After all, an issue like this is quite a common headache for programmers, so much so that there is a well-known short verse in which a computer owner laments about wanting to sell their machine on the account of it not doing what the programmer wants it to do, but only what it is literally told to do (Rawlins, 1997, p. 80).

From Croesus to Computers **85**

But suppose that something similar may happen with an automated system that operates something that carries a greater degree of responsibility, or a system that has overall greater capabilities than a small gaming algorithm. Imagine a network of self-driving cars with an imperative to minimize the occurrence of car crashes. Suppose also that this network is capable of parsing natural language and also capable of learning by observing. After seeing how two pedestrians technically never encountered a car crash with one another, the network might decide that the safest method of navigating a city is to drive at walking speed. In an even more speculative vein, imagine we give an open mandate to an artificially superintelligent agent to act as a world administrator and give it instructions to ensure humans can live out their lives peacefully and without worries, whereupon the agent decides to fill the atmosphere with a neurotoxin designed particularly to disrupt nerve connections responsible for unpleasant emotions. A particularly treacherous hypothetical would be the one in which we are trying to instil a basic notion of morality into an autonomously acting AI specifically to prevent harmful behaviour. We might want to do that by showing the AI numerous examples of people acting altruistically and in a self-sacrificing manner. However, the AI might interpret that data in a sense that self-sacrificing humans are expendable, ending up with a completely twisted ethics from the intended.

In all of these scenarios, artificially intelligent agents have figured out unorthodox methods of working with the instructions they were initially given. Arguably, us humans would not only consider these results as invalid, but we might find them downright harmful. Perhaps even so harmful that they would rightfully constitute an existential risk for humanity, a possibility to end humankind. And the more power an agent has at its disposal, the more dire these solutions might end up being. These scenarios in which artificially intelligent agents use an unexpected method for reaching their goals, especially if these methods have disastrous consequences, are often called *perverse instantiations.*

The problem of perverse instantiations is mostly associated with very capable agents such as artificial general intelligences (AGI) or even superintelligences and is an important topic in the fields of AI safety and ethics. Significant research has already been done on both speculative and factual AI failure scenarios (Yampolskiy, 2018), but more are always welcome. Granted, the most dire scenarios that involve an existential risk are likely long ways away in the future, if they are at all possible. There are some unrepentant optimists who find the AI doomsday scenarios ultimately unlikely (Pinker, 2018). More moderate optimists acknowledge that a risk exists but believe that the solution will be well within our grasp (Agar, 2016). On the other hand, the most dreadful scenarios need to happen but once.

It should be noted, however, that these fictional cautionary tales did not first occur in the minds of contemporary computer scientists and philosophers of AI. There are plenty of tales from the past in which the carelessness or hubris of its protagonists brings about something similar to a perverse instantiation event. Genies that interpret commands too literally, a monkey's paw that adds malicious consequences to wishes, even misunderstood prophecies can have significant parallels to perverse instantiation scenarios. Researchers and commentators often

86 Goran Rujević

point out these similarities (Bossman, 2016) and some even use them as sources of valuable analogies in their research (Yudkowsky, 2011).

There is a vast variety of cautionary tales of this kind from all over the world, from folk tales to theatre tragedies. Some embody the literal moral of 'beware what you wish for' and show how wishes that are not thoroughly thought through can have dire consequences, as is the case with the story of king Midas. Others are more nuanced and include misunderstandings or badly interpreted prophecies, as is the case with king Croesus or Macbeth. It would seem that humanity struggled with a kind of perverse instantiation problem since our earliest days, of course, usually tied with the cases of misunderstandings between humans or between humans and hypothetical supernatural powers. All of these stories, ancient and contemporary alike, belie a problem that is fundamentally philosophical in nature, a problem so old that even Socrates had to face it in his dealings with the treacherous sophists.

The aim of this article is thus to uncover this fundamental problem and show its connection to the contemporary field of AI ethics. In doing so, we cannot hope to definitively solve the problem of perverse instantiation, but we can demonstrate that the transdisciplinary dialogue between logic, philosophy and computer science holds the best promise of finding a solution. In order to achieve that, we will first examine the idea of perverse instantiation in more detail; then we will reflect on the common thread found in cautionary tales that deal with this theme and demonstrate how it is related to philosophical investigations of the Socratic kind; afterwards, we will try to draw further parallels with the contemporary concept of perverse instantiation and finally reflect on the implications this may have for future investigations.

Twisting Wishes and Instantiating Perversely

The most exhaustive account of perverse instantiation was penned by Swedish philosopher Nick Bostrom in his seminal book *Superintelligence; Paths, Dangers, Strategies* (Bostrom, 2014). There, it is considered as a kind of 'malignant failure mode' of development of a (machine) superintelligence. A failure mode is any of the myriad ways an attempt at creating a superintelligence can end up being unsuccessful. Most of the conceivable failure modes are 'benign' insofar as they do not carry with themselves an existential risk. Certainly, benign failure modes can be very impactful and can range in anything from technical mishaps to financial troubles, but they are at worst only fatal for the specific developmental project in question and do not pose a general risk for mankind. On the other hand, malignant failure modes have an impact far beyond the specific project and can spell doom for the whole human race. This may sound dramatic and malignant failure modes may themselves be highly unlikely, but when the future of humanity is at stake, such failures only need to happen once, and Bostrom even points out that an important feature of malignant failures is that they eliminate the option to try again (Bostrom, 2014, p. 146).

From Croesus to Computers **87**

In Bostrom's words, perverted instantiation is 'a superintelligence discovering some way of satisfying the criteria of its final goal that violates the intentions of the programmers who defined the goal'. (Bostrom, 2014, p. 146). A now already well-known thought experiment illustrating such a case is one in which a super-intelligence decides to fulfil the given goal 'Make us smile' by paralysing facial muscles of every human being into a rictus grin. Technically, the goal is achieved, except that we can be pretty sure that the principals (programmers who gave this goal to the superintelligence) definitely did not have this result in mind.

At first glance, it is easy to recognize that the command 'Make us smile' is terribly articulated. Tentatively, the principals wanted the superintelligence to make everyone in the world happy, but the above goal is merely a roundabout way of expressing that wish. Moreover, the goal 'Make us smile' is rather vague, and some might think that this vagueness of the concept of smiling is what directly contributed to the fatal misunderstanding between man and machine. Clearly, the solution would be to deliver a command that will explicitly forbid the superintelligence from taking the twisted approach. But Bostrom anticipated such a response and showed that a command 'Make us smile without directly interfering with our facial muscles' can lead the superintelligence to interfere with our muscles only indirectly, via motor centres in our brains (Bostrom, 2014, pp. 146–147).

Perhaps setting explicit limitations in a command is not a good approach. After all, there are many ways in which a command can be misunderstood and pre-emptively eliminating every single one of them particularly is not a feasible approach. Not to mention that a superintelligence would certainly be able to find ever more innovative ways of misinterpreting a command than the smartest of humans could. A better approach might be to lean into the superintelligence and give it a goal via a more complex command, hoping that it would be able to parse it in adequate manner. Maybe we ought to explicitly instruct the superintelligence to 'Make us happy'.

Bostrom notes that this command is as easily twisted as the last. The super-intelligence might conclude that accounting for idiosyncratic ways each human can be made happy is simply too complex and that a more economical approach would be to implant electrodes into everyone's brains that would specifically stimulate the pleasure centres. Or it might take even greater liberties and create digital copies of human minds that can then be directly modified into the state of eternal bliss. In one way or another, these actions fulfil the given goal of happiness (Bostrom, 2014, p. 147).

However, if the superintelligent agent is, indeed, superintelligent, why do we expect it to behave as if it is absolutely ignorant of the intentions behind the commands it is given? Humans of quite average intelligence can clearly parse what is actually meant by both 'Make us smile' and 'Make us happy' commands, a superintelligence should be able to do at least the same, if not better than that. Bostrom offers a chilling reply why a superintelligence would still choose to twist the commands: 'The AI may indeed understand that this is not what we meant. However, its final goal is to make us happy, not to do what the programmers meant when they wrote the code that represents this goal' (Bostrom, 2014, p. 147).

88   Goran Rujević

If a superintelligent AI presents such significant danger of becoming an instrumental fiend, perhaps we should be arduously working in the direction of furnishing our future artificially intelligent agents with an internal mechanism that would guarantee its compliance with well-being of humans. We ought to teach it specific ethics or give it a kind of conscience, so to speak. After all, it is conscience and a system of laws that is promoting agreeable behaviour among humans. Sadly, there is no guarantee that this would work with a superintelligence as any additional system for conscience can also be subverted: either removed so it does not interfere with regular instrumentality or reconfigured such that it is rendered ineffectual (Bostrom, 2014, p. 148).

These hypothetical scenarios are not a product of mere pessimistic thinking about the worst possible cases. Bostrom explains that malignant failure modes of superintelligence development are a reasonably expected result of three other points he established earlier in his book. First is that a superintelligence can achieve a position so dominant and influential that it can affect humanity as a whole (see Bostrom, 2014, p. 109). Second point is Bostrom's well-known orthogonality thesis which states: 'Intelligence and final goals are orthogonal: more or less any level of intelligence could in principle be combined with more or less any final goal'. (Bostrom, 2014, p. 130). In other words, we should not expect that a superintelligence will have goals that are aligned with ours or that it will approach its goals in ways we tend to. Third point is the so-called instrumental convergence thesis which, to put it shortly, states that a certain goal can be seen in a convergence of multitude of instrumental values, some more general than others, which the agent will then be likely to pursue whether or not the principal had them specifically in mind while articulating the command. This might lead the agent to perform some unexpected actions, such as ensuring that no one can turn it off or change its goals (Bostrom, 2014, pp. 132–133). All of these points taken together lead us to the picture of a capable agent with possibly alien goals and values that might not align with ours – a perfect storm for a malignant failure.

In order to avoid this and many other pitfalls of artificial superintelligence, we can either limit the executive capabilities of the AI (for instance, by vetting any of its actions, by constantly monitoring for signs of trouble, or simply by 'boxing' the AI into an isolated environment), or we can work on making sure the AI works with goals and motives that are aligned with those of humanity at large (Bostrom, 2014, p. 157). Limiting the efficacy of the AI is a crude method and some might say it defeats the purpose of an autonomous AI agent. The other option is more far-reaching, but also more difficult to implement, as it effectively means teaching morality to an artificially superintelligent agent. This, of course, can't be as simple as Asimov's laws, which, as it turns out, require quite a lot of work to be formalized (Weld & Etzioni, 2009). Should we teach AI morality from principles or from examples? And how can we make sure our ethics lessons are themselves received properly and not in any way perverted?

An interesting solution to this problem can be found in something Eliezer Yudkowsky called *Coherent Extrapolated Volition* (CEV). In his words:

From Croesus to Computers **89**

> In poetic terms, our coherent extrapolated volition is our wish if we knew more, thought faster, were more the people we wished we were, had grown up farther together; where the extrapolation converges rather than diverges, where our wishes cohere rather than interfere; extrapolated as we wish that extrapolated, interpreted as we wish that interpreted.
>
> (Yudkowsky, 2004, p. 6)

The idea of Coherent Extrapolated Volition, if implemented properly and on a sufficiently intelligent AI will make it so that the AI cannot now take our wishes or commands at face value, but must interpret them as we would interpret them in the most beneficial of circumstances. The more intelligent an AI is, the better it will be at CEV. This effectively nullifies Bostrom's scenario in which an AI may well understand what we want, but doesn't bother to take that into account.

There are a couple of caveats with CEV. First of all, it doesn't so much solve the problem of misaligned goals and values as it 'outsources' its solution to the very AI in question. Second, implementing the CEV will require a very powerful AI, perhaps even a superintelligent one, as it will not only have to model our hypothetical wants, but also model them in very specific and optimal circumstances. Surely, we cannot wait for the emergence of superintelligence to be able to solve safety issues that arise with superintelligence, that would be far too risky. Is there something we can do in the meantime? After all, we can have AI failures before we have superintelligences.

It should be noted that the scenarios Bostrom described with regards to perverse instantiation always deal with a superintelligence. We might want to generalize perverse instantiation so that it encompasses less advanced systems, but in that case, we need to make a few modifications to the concept. First of all, simpler AIs represent comparatively smaller risks, probably far from an existential risk. Original perverse instantiation is considered a malignant failure mode, that is, linked to a significant existential risk. However, even non-existential risks can be significantly impactful and may lead to unwanted results that are irreversible. And even if we end up talking about mere inconveniences here, a prudent attitude towards developing AIs should take even minor gripes into consideration.

Second, a simpler AI probably will not possess the ability of self-editing that was central to Bostrom's argument against programmed conscience. It might be that simpler AIs could be reigned in with artificial conscience. On the other hand, a simpler AI would be comparatively less competent at interpreting goals and as such may present a greater risk of twisting the letter of the command. With that in mind, the risk of perverse instantiation with simpler AIs is not simpler per se, nor should we think of it as any less relevant, it is merely less dramatic. It is still worth paying attention to.

Finally, we ought to make a remark on a specific terminology we have been using on preceding pages. A careful reader might have noticed that we used terms 'goals' and 'commands' almost interchangeably. One might argue that there is a difference between them insofar as we can consider 'commands' to be specific

90 Goran Rujević

articulations used for expressing and communicating goals, whereas 'goals' are desired end-states towards which a certain endeavour (of humans or of machines) is directed. However, for the point we are trying to make in this article, it is important to recognize that a specific articulation of a command can profoundly affect what goal will be pursued. Arguably, this is the difference between the 'Make us smile' and 'Make us happy' commands which in the minds of the programmers communicated the same goal, but for the superintelligent agent these stipulate different goals. In that respect, we should keep in mind that a perverse instantiation can result not only from a specific goal, but also from a specific command used to issue it.

Aligning Intensions and Extensions

Gottlob Frege explained in his famous article *On Sense and Reference* (Frege, 1892) that the meaning of a sign is no simple matter, and that we can separately talk about the *sense* of a sign (mode of presentation) and the *reference* that sign makes (the definite object). Today, we are more likely to use the terms *intension/ connotation* and *extension/denotation* of concepts. By intension of a concept we understand all descriptors that said concept represents as a shorthand, whereas by extension of a concept we understand all entities that would fall under the umbrella of that concept. The intension of the concept 'triangle' would be 'a geometric figure with three sides and three angles' and its extension would be individual triangles of all possible sizes and configurations. Intensions and extensions of concepts can be communicated via *definitions*, and correspondingly we have intensional definitions (usually descriptive) and extensional definitions (most often enumerative). Some definitions are good, appropriate, some are poor and imprecise. As with any attempts at communication, definitions can be more or less successful. When two parties have the same understanding of intension and/or extension of a concept, we can say that their perceived intensions and/or extensions are aligned; when two parties are disagreeing or misunderstanding, we can say the perceived intensions and/or extensions are misaligned.

Normally, intension and extension are associated with concepts and terms, most of which are represented by one or a few words: Greeks, mortals, triangles, Morning Star, farmers who own donkeys, et cetera. But if we can assign intension and extension to short phrases, why not also to longer ones, perhaps even larger elements of discourse. We would like to propose that even groups of sentences can have intension and extension, coherency of which only depends on if the group has a common subject. In other words, we can generalize the ideas of intension and extension to many different discursive forms. One can say that, for instance, a sonnet about the beauty of one's beloved has both intension (the descriptions) and extension (the beloved). This goes for prophecies, too, and is even useful in understanding them properly as they often tend to be vague, cryptic, potentially even misleading. Intension of a prophecy is the concrete message delivered, opaque as it may be; extension of a prophecy is the specific future event that it refers to. Prophecies can be useful plot elements in a story as they can serve as a

From Croesus to Computers **91**

source of conflict, especially when they are wrongly interpreted, that is to say, when the subject of the prophecy and the prophet experience misalignment of intension, extension or both.

The classical tale of Croesus is a good illustration of this. As the story goes, Croesus, the king of Lydia, was preparing to go to war with the Persians. Like any pious man of his time, Croesus was aware that the gods can be meddlesome and capricious, and that he should consult the Oracle before any important decision. When he asked if it was an auspicious moment to attack the Persians, the prophetess responded in an expectedly roundabout manner: 'If Croesus should make war on the Persians, he would destroy a mighty empire' (Herodotus, 1904, p. 19). Emboldened by this prophecy, Croesus marched against the Persians, his army ended up being soundly defeated which led to the destruction of the mighty kingdom of Lydia. In this tale, both Croesus and the prophetess likely fully understood what the concept 'mighty empire' means, but there was a discrepancy with regards to what this concept can refer to: Croesus never stopped to consider that the extension of that concept could include his own kingdom.

In Shakespeare's *Macbeth* we find an even more complex instance of misunderstanding, again in the form of a poorly interpreted prophecy (Shakespeare, 2003). When Macbeth visits the three witches in Act 4, they conjure apparitions that give hints about the events to come. One of them implores that Macbeth should beware his rival Macduff while another delivers a prophecy that none of woman born shall harm Macbeth. While previously wary of Macduff, Macbeth is now relieved as he thinks he needs not fear Macduff nor anyone else because the prophecy stated that no one can harm him, reasoning that everyone has to have had a mother at least once in their life. However, it later turns out that Macduff can, in fact, harm Macbeth due to a 'technicality' that Macduff was born by a Caesarean and not by natural means. There are multiple layers of misunderstanding the phrase 'none of woman born' here. First, Macbeth mistakenly believes that the extension of this phrase is literally no one, an empty set, but the true extension of the phrase was at least Macduff. It is irrelevant if Macbeth knew of the circumstances of how Macduff came into this world, for Macbeth also misunderstood the intension of that phrase – he believed that being born of a woman means having a mother and everyone must have a mother; however, the apparition that delivered the prophecy seems to think that being born of a woman only counts in cases of natural birth and that babies brought forth by a Caesarean are not technically born but ripped out of the womb. Not only was Macbeth foolhardy to dismiss the first warning about Macduff, he also ended up missing both the intension and the extension of the prophetic phrase, which ultimately led him to his demise.

Fictional world is not the only place where this conceptual misalignment might occur. The real world is home to such cases as far back as ancient times. In ancient Greece, the sophists were known to be clever and sharp-tongued individuals, skilled in law and rhetoric, for which they offered their service as private tutors, mostly to wealthy citizens of the poleis who wished to become more skilled in public discussions, political debates and legal proceedings. Yet, for all their skill in these areas, sophists were often regarded with a certain dose of disdain because

92   Goran Rujević

they employed these skills for profit and not for pursuit of truth. Sophists were notorious for sometimes twisting truth and employing dishonest tricks in a debate just so that they could emerge victorious. These tricks are therefore known as *sophisms*.

One of the best known sophisms is the one about a veiled person. A sophist brings out a person concealed beneath a veil before the crowd and engages a particular person in the crowd with the following questions: 'Do you know your father? Yes. Do you know this man who is veiled? No. Then you do not know your father; for it is your father who is veiled'. (Enfield, 1837: 111). Admittedly, when compared to some sophisms that are truly clever, this one does not seem very highbrow as it shows the sophist seemingly outwitting their interlocutor on a technicality. However, this technicality is rather relevant to our investigation as the sophist's trick can be said to be one of an underhanded shift of a concept: when the sophist first asks if the interlocutor knows their father, the term 'know' is likely to be interpreted as 'having previously met'; but in the second question, the term 'know' is more likely to be interpreted as 'able to identify'. These are two similar, but still different intensions of the term that then lead to different extensions, yet the sophist afterwards completely equates these two uses of a term. We might say that at the root of this sophism is a misalignment of both intensions and extensions of the concept 'knowing' between the sophist and their conversational partner.

The philosophical method of Socrates, who was a contemporary of many sophists, was directed precisely against this kind of obfuscation of meaning. In contrast with the sophists who touted their own knowledge and wit, Socrates famously claimed to know nothing and used that as his starting position in the search for truth. Rather than accepting given or popular claims, Socrates first examines if the presuppositions from which a discussion starts are at all universally acceptable. Before any specific examination is undertaken, Socrates first seeks the very nature of the concept that is in question. Before pontificating about anything, we need to have transparent definitions precisely so that we can avoid situations in which we think we know but we don't, or situations in which we think our understandings are aligned but they are not.

Quite often in dialogs written by Plato (see Plato, 1997), Socrates is in direct conflict with sophists, engaging them in a dispute and often showing that their knowledge is lacking. In *Protagoras*, Socrates famously asks the eponymous sophist to try to keep his answers clear and succinct and not resort to rhetorical ramblings as a way to avoid discussion. In *Greater Hippias*, Socrates rejects the sophist's attempts to merely give examples of beautiful things and insists on finding the definition of beauty itself. In *Gorgias*, Socrates engages in a long sequence of questioning in order to come to an agreement with the sophist on what the art of rhetoric truly is. One could say that Socrates worked hard to avoid cases of misalignment in understanding of intension and extension of important concepts.

It should be noted that the method of Socrates very rarely ends with a positive result. The participants of his dialogs often leave with more new questions than answers and we can rarely settle on any single definition because of frequent exceptions and limiting cases. And yet, we can say that the participants of a dialogue with Socrates leave such encounters a little wiser than they were before – if nothing else, at least they are aware of shortcomings in their understanding of things. To a certain extent, this is emblematic of most philosophical undertakings, which goes to show that philosophical inquiry is never a finished business, but rather a perennially open project of refinement.

Genies Once and Genies in the Future

What is, then, the point of intersection between the two fields described above? What do Socrates and the sophists, Croesus and Macbeth have to do with artificially intelligent agents and perverse instantiation? We argue that much like Socrates was trying to eliminate ambiguity from his dealings with the sophists and reach true understanding of concepts, so do principals seek to formulate unambiguous instructions for artificially intelligent agents in order to reach true alignment of goals and avoid perverse instantiation.

Furthermore, just like we generalized intension and extension from concepts to entire prophecies, a similar generalization can be done from concepts to wishes or commands a principal gives to an agent. In this, we follow the idea that perverse instantiation is a complex occurrence that can arise from different types of failures (Aliman & Kester, 2019). An intension of a concept is akin to the contents of the given command, the notion behind it; much like the intension contains general descriptors associated with a concept, the command can also be thought of as description for a certain goal. For instance, the command 'I wish to become the wealthiest person on Earth' has the intension 'I want you to arrange events in such a way that, compared with all other people now existing on Earth, I own the most assets'. On the other hand, the extension for a concept can be seen as possible methods for reaching that goal; much like extension encompasses all concrete objects that fall under a concept, so do methods encompass concrete steps, algorithms and heuristics that are to be undertaken in order to realize a goal. The principal issuing the previously mentioned command rightly assumes that an extension of that command is that the genie/agent would furnish them with new wealth that exceeds the wealth of all other people on Earth; however, the method that also falls under the extension of that wish is that the agent could turn every other person on the planet into a pauper, making the principal the richest person not by giving new wealth but by taking wealth away from everyone else. As with any concept and prophecy, both intensions and extensions of wishes/commands can be well understood or misunderstood, aligned or misaligned between the principal and the agent.

Formulating a command with minimal or no chance of being twisted by the agent is akin to having a clear and distinct idea (in a Cartesian context) or to having a concept with well-formulated intensional and extensional definitions. Unfortunately, arriving at definitions that are perfect and not prone to misinterpretation and disagreement is very difficult. That much was evident from the results of Socrates' efforts which seldom produced a definitive definition of a

94 *Goran Rujević*

concept. The same can, then, be expected of commands that would be issued to a certain agent, whether it is a human or an artificially intelligent agent. Poorly-articulated commands are like imprecise definitions, and perverse instantiations are like misunderstood definitions. Even our immediate responses to such imaginary scenarios are similar: when faced with shortcomings of their initial definitions, Socrates' interlocutors usually try to provide additional stipulations to their first attempts; as Bostrom predicts, one of the first attempts to avoid perverse instantiation is to add stipulations to the command that would rule out undesirable outcomes (Bostrom, 2014, p. 146). As valiant as these attempts may be, they do not guarantee that new definitions/commands will be good, that is, they do not guarantee that they will enable perfect understanding or alignment.

With regards to mutual understanding of concepts between people, the intensional and extensional dimension are not orthogonal or independent from one another, but also not fully correlated. An alignment in understanding of a concept's intension is likely to produce and alignment in understanding its extension, though not always. If two people agree upon a detailed and exhaustive description of a concept, it is reasonable to assume that they will also agree upon whether a certain object corresponds to this description or not, meaning that it falls under that concept or does not: if two people work with the same definition of a square, they will likely sort the same geometric figures under squares and non-squares. A misalignment of extensions can still occur with aligned intensions, however, particularly if the intensional content is vague or there happens to be an error in perception of certain objects. For instance, two people can agree on what a mammal is, yet one of them can mistakenly infer that platypodes cannot be mammals because they lay eggs; the error here is one of incomplete knowledge, as the second person was not aware that platypodes breastfeed their young.

The reverse case is slightly different. An alignment in understanding of a concept's extension only sometimes produces alignment in understanding its intension because inferring intension even from a complete list of extension requires serious capabilities of abstraction and pattern recognition. Even if two people are equally capable of finding a common thread within a sample of objects, there is no guarantee that they will find the exact same common thread. For instance, if we give two subjects a set of photos of 15 blue-eyed people, one of them might infer that this is a group of blue-eyed people, while the other might infer that this is a group of people born after photography was invented. Both are strictly speaking correct, but they still arrived at distinctly different intensional descriptions of the given group.

On a slight side note, this asymmetry in the reliability of inferring extension from intension and vice versa can be considered a fundamental reason why intensional definitions of concepts (ones that describe their contents) are often more highly valued than extensional definitions (exhaustive lists). This explains why Socrates scolded Hippias not to enumerate examples of beautiful things but rather provide a definition for what beauty is, as well as why we generally prefer intensional over extensional definitions.

For us in this moment, the relevant insight is that despite the link between intension and extension of a concept, there can be cases in which there is alignment in understanding one and misalignment in understanding the other. Therefore, there are four possible situations of how two parties can understand a concept:

a. both intension and extension are aligned between them (complete understanding)
b. there is alignment of intension but misalignment of extension
c. there is misalignment of intension but alignment of extension
d. both intension and extension are misaligned (complete misunderstanding)

We would now like to argue that for every combination except for a) there is a representative cautionary tale *and* a corresponding speculative perverse instantiation scenario. Combination a) is the obvious exception here because complete understanding lacks any tension and drama and would thus make for a poor plot of a tale and an utterly uninteresting premise for a speculative scenario.

Combination b) represents the classical case of tragic misunderstanding: both parties agree upon what is said, but disagree upon what that refers to. In a wish-granting context, this case is well embodied in the story of the Prague Golem. In one of its versions, Rabbi Loew created the Golem to perform various laborious tasks. One time, the Rabbi instructed the Golem to fetch water from the well, a task that the Golem carried out successfully over and over again, until the Rabbi's house was flooded. The Golem understood the letter of the command, it and the Rabbi were in good intensional alignment. The issue arises in that the Rabbi assumed that the Golem would fetch but one bucket of water, which is not how the Golem approached the task – an obvious misalignment of extension. Of course, the entire conundrum would have been avoided had the Rabbi instructed the Golem to bring *one* bucket of water from the well.

A perverse instantiation that would fit this bill is the one that would stem from poorly worded commands. Imagine a hypothetical future corporation which obtains a clever general artificial intelligence with the intent of using it to guard their sensitive trade secrets. The AGI is issued a command 'Find a definitive way that will stop our rivals from attempting to steal our secrets'. It then might reasons that the rivals are only trying to steal the secrets because they do not possess them and if they did possess the secrets they would no longer be trying to steal them. With such reasoning, the AI decides to deliver the trade secrets to all rivals of the company. In this case, both the company and the AI well understood what words of the command say (aligned intension), but there is a disagreement with regards to what actions fall under the extension of that command (misaligned extension). Had the command been worded in a more careful manner, like 'Keep our secrets out of our rivals' hands', this perverted instantiation wouldn't have occurred. But a different one still might – for instance, the AI might conclude the best way to do that is to eliminate the rivals via hostile takeovers or sabotage. On second thought, this might not be a perverted instantiation as it is pretty much how companies already operate today.

96   Goran Rujević

Combination c) is an interesting case insofar as it can occur in two different variations: a malicious and a benign one. The malicious variation is any of the many 'monkey's paw' stories where a person utters a wish with a certain beneficial idea behind it (intension), the wish-granter interprets the wish in a twisted way (misaligned intension) that does grant the wisher what they wanted (aligned extension) but also something unwanted in addition that is the result of misaligned intension. In the classical story by William Jacobs, when Mr White wished for 200 pounds, he did in fact get what he wished for, but in the form of the settlement for the death of his son (Jacobs, 1910). The death was obviously not part of the wish, but the monkey's paw interpreted the wish most maliciously. Unlike the previous combination, the issue here is not that the wish was hastily worded. It is hard to think of a way in which Mr White could have worded the wish that would preclude any and all unwanted side effects. It is more likely that the very act of wishing was not thoroughly thought-through.

With perverse instantiations, it is clear how an intentionally malicious artificially intelligent agent would be harmful. It would practically be the monkey's paw, guaranteeing that any command it is issued will go wrong. But let us suppose humanity will never produce a patently evil artificial intelligence or, at the very least, will not use one that is demonstrated to be such. There is still a possibility of this kind of misalignment even with non-malicious agents. Arguably, Murphy's *playfun* algorithm was one such case: the concept of 'winning' in *Tetris* is difficult, if not impossible to properly articulate, leading the algorithm to a state which can technically be at least called 'not losing', albeit at the cost of not playing *Tetris* at all. The more we think about it the less surprising this event is, because how can one win at a game that erases successes (complete lines) and preserves mistakes (incomplete lines).

The benign variation of the combination c) would be one in which the wish-granter is given a poorly thought-out wish, yet they are benevolent enough to look past that and interpret the wish in a slightly different way (intension misaligned, but for the better) such that the wisher receives the best results they wanted. There are not many cautionary tales describing such cases, possibly because there is nothing to be cautious about when there is a powerful and benevolent wish-granter around. In the context of AI, this is Yudkowsky's friendly AI working on Coherent Extrapolated Volition: interpreting our words with our best interests at heart.

Finally, combination d) is a complete breakdown of sensible communication. These are not just innocent misunderstandings, these are complete follies where the wishing party delivers a poor request and receives even poorer results. A good illustration of this would be the ancient myth of goddess Eos and her mortal lover Tithonus. Wanting to spend the eternity with her lover, Eos implored Zeus to grant Tithonus immortality, which Zeus promptly granted. In other accounts it is Tithonus himself who asked for this boon. However, in wishing for immortality, the wisher neglected to ask for eternal youth. The ever-ageing but never-dying Tithonus ultimately shrivelled down into a cicada. This wish was both foolhardy, as it asked for something mortals by definition shouldn't have, and reckless, as the wisher hadn't properly considered that eternal life without eternal youth would be

From Croesus to Computers   97

nightmarish. Effectively, the wish of immortality was a poor way to convey what the wisher actually wanted, causing a misalignment of intension, and the result of the wish is definitely not what the wisher wanted to happen (a misalignment of extension).

A perverse instantiation corresponding to this combination would be the most malignant of all, but also least likely to occur. Bostrom's example with an artificial superintelligence that is given the command to 'Make us smile' is one such scenario. First, the command is poorly-worded: the principal issuing it probably equates smiling with happiness and disregards that a rictus grin might also be understood as smiling (misaligned intensions of the concept 'smile'). Second, the results of the command is something the principal definitely did not want (misaligned extension of the command).

One can open a debate on how many of these examples are a result of intentional malice, but that would be beside the point we are trying to make. The point is that misunderstanding wishes or commands is in principle possible due to limitations of communicating intensions and extensions of ideas between different parties, benevolence or malevolence only adjust the probability of (mis)understanding, but can never eliminate them outright. So, to Pinker and other optimists who claim that there is little reason to think future AIs will be malicious one can respond that these issues can arise regardless of a machine's conative stance. As long as there is a possibility of a certain scenario playing out and we are aware of that, it should not be controversial that we would want to take it into account.

An objection can be made to this general analogy that the idea of intensionality does not translate well to the field of artificial intelligence. Intension of a concept is, in a sense, rather personal and subjective, and trying to find out how a machine 'actually understands' the content of an instruction might seem a futile effort. An AI based on neural networks may end up being so complex that not even its creators can be absolutely certain how it functions – it becomes a veritable 'black box'. What hope is there to talk about the kind of intension that AI harbours? Even more generally, both intension and extension make sense in the context of human consciousness, and the question of machine consciousness is still a hotbed of debate. Bostrom specifically notes in *Superintelligence* that his ideas do not rely on ascertaining AI subjectivity (Bostrom, 2014, p. 338). Introducing analogies with human subjectivity would then seem to be a step in the wrong direction.

However, this objection is merely superficial. First, the notion of intensionality has already encountered private subjectivity among humans without much problem. We amongst ourselves cannot 'truly' know how one 'actually understands' the content of a concept (at least not until brain imaging and neurosciences become sufficiently advanced). All we can rely on is the not-so-perfect tool, the language. So far, it has proven itself useful enough. In a similar way, we have something similar to language with machines, that being the code they are executing, which can keep this analogy going. Second, machine consciousness is not a prerequisite for our analogy. Much like we can talk about 'machine learning' and 'memory' without implying that there is a consciousness that learns and remembers, we can also talk about 'understanding' or '(mis)aligning'

98   *Goran Rujević*

intensions and extensions in a metaphorical way. Furthermore, the 'black box' of machine learning is not necessarily completely opaque nor utterly alien to our understanding. Researchers working on *AlphaZero* chess-playing neural network have found that the network contains representations of concepts that are similar to what exists among human players (McGrath et al., 2021). Not knowing precisely how something works is not the same as being utterly ignorant of it.

Legacy of Socrates

The comparison done on previous pages shows in no uncertain terms that there are significant parallels between traditional tales warning us about dangers of reckless wishes and hypothetical scenarios of perverse instantiation. But these similarities were already well-known. Our analysis, however, demonstrates that the similarities can be conceptualized as a fundamental problem of misalignment of conceptual intensions and extensions, a problem that is ultimately philosophical, logical in nature. This insight can now point us in the direction of possible approaches that can be taken to solve these problems, at least until Coherent Extrapolated Volition becomes a viable prospect.

Misalignment of intension or extension of terms is a frequent occurrence. To combat it, thinkers of all ages quite naturally resorted to establishing definitions. That is exactly what Socrates tried to do in his clashes with the sophists – instead of allowing them to twist and change meanings of terms at will, he insisted on defining base terms before any further debate can continue. Defining a term can be understood as an act of 'fixating' its meaning: in a large possibility space of meaning of a term, the act of defining singles out a specific one. As we already mentioned, both intension and extension of a term can be defined, giving us intensional definitions and extensional definitions. Intensional definitions stipulate what a term is, perhaps by offering the *genus proximus* and *differentia specifica*, whereas extensional definitions stipulate what objects fall under the term being defined, which is usually done by some kind of enumeration of these objects. In colloquial dealings, we use definitions to prevent misalignment in understanding the meaning of terms. If Croesus could have gone over the definitions relating to the prophecy from the Oracle, he might have realized that the term 'mighty empire' can refer to his own. If Mr White was given an exhaustive enumeration of all ways in which he would earn 200 pounds, learning that among them was his son's death would likely have prevented him from ever using the monkey's paw. Most of the time an intensional definition will be sufficient, sometimes an extensional definition is better, but the best results are evidently ones in which both types of definitions are used.

If we now shift into the realm of artificially intelligent machines, we will see that we have similar options. In order to ensure the AI agent understands our commands properly, we need to teach it about how we expect it to act when faced with future commands, and we can do it either by offering general instructions or by providing it with numerous examples. In essence, this is similar to the problem of instructing ethical behaviour to machines. We can do that in the classical paradigm of writing general instructions articulated in such a way that the machine will be capable of associating specific commands with stipulations/ limitations from high-order instructions – the so-called 'top-down approach' in which machines are given high-order ethical principles to be used in specific circumstances (Wallach et al., 2008: 568–569). This would be analogous to providing the machine with an intensional definition of moral action. On the other hand, we can rely on the newer machine learning paradigm in which we present the machine with a vast number of specific examples of what we deem to be acceptable behaviour and expect the machine to formulate an adequate model based on these examples. This would be the so-called 'bottom-up' approach (Song & Yeung, 2022), which would be analogous with providing the machine with an extensional definition of morality.

When we compare the two principal approaches to machine learning with the two types of definitions, even more similarities between these areas crop up. The challenges of 'top-down' approach include the challenge of consistency within programmed principles, then the issue of selecting an appropriate ethical framework and the issue of properly articulating ethical principles in a format accessible to an AI, just to name a few. A similar set of issues are known to arise around intensional definitions. First of all, not all concepts are equally easy/hard to define and some definitions may end up in conflict with others. Formal concepts like 'triangle' or 'bijection' tend to fare far better in that regard than more diffuse concepts like 'justice' or 'piety'. It can be very hard to set forth a definition such that it is both inclusive enough to allow for outlying cases to be encompassed and at the same time restrictive enough not to allow for potentially confounding intruders to slip in. For instance, how can we properly define a 'chair'? At first glance it may seem easy enough to say that it is 'a piece of furniture that is used for sitting', but would that mean sofas and couches are also chairs? What about tables, some people sit on those, too? And tiny model chairs, are they not chairs even if we cannot sit on them? If we imagine this kind of edge cases with regards to moral concepts, it becomes evident that they can potentially lead to perverse instantiation.

Extensional definitions and the 'bottom-up' approach also share a number of issues. Can we be sure that a collected set of training data for a machine is adequately representative of the morality we are trying to illustrate? And even if it is, can we be sure that the AI will extrapolate from it useful guidelines for future action? Will these guidelines be something others (humans) find agreeable? In a similar vein, it is hard to know if an extensional definition is exhaustive enough to determine the relevant concept, although the large number of individual examples a machine can process can help with that. Then again, there is no guarantee that that even a massive set of examples is not exemplifying some other common characteristic. This was famously pointed out by Ludwig Wittgenstein when he claimed that ostensive definitions, a kind of extensional definition in which we define a term by pointing at a single exemplar of it, are notoriously unreliable (Wittgenstein, 2009, p. 17). If we meet an alien who is trying to learn our language and attempt to introduce them to the concept of a 'brick' by pointing at one, the alien will have no way of knowing if we are referring to that particular object or to

100 Goran Rujević

the class of that object in general; it would also be unclear if we are pointing at the object itself or merely at its shape, colour, position or something else entirely.

AI ethics is not the only field where these considerations are relevant. The currently dominant paradigm of machine learning that relies on processing vast amounts of training data is akin to learning by studying very exhaustive extensional definitions. To expand our Socratic metaphor, contemporary machine learning has the capacity to produce very *sophist*icated systems. While we may think that machines have a distinctive advantage over Socrates and Hippias in that they can process a lot more examples in far shorter times than humans ever could, relying exclusively on extensive extensional defining is still prone to errors of edge cases, which might lead to perverse instantiations in the future. Murphy's *playfun* algorithm is just a simple illustration of how using only exemplary training data can create unexpected outcomes. More complex examples exist, for instance, Goodfellow et al. have shown that adversarial examples can quite easily subvert machine learning models (Goodfellow et al., 2014).

With all of this in mind, we believe that a good inspiration for our future actions, especially in the field of AI ethics, can be found in the figure of Socrates. First, his insistence on determining what is the nature of the thing he is discussing with his interlocutor is, as we have already pointed out, a call to find a proper intensional definition. Even if such a call turns out to be at the moment too difficult or even futile, the mere attempt at reaching it still provides Socrates and his listeners with new insights. Second, although Socrates scolds his conversational partners when they offer extensional definitions in place of intensional ones, he is still not altogether averse from examining individual examples in order to arrive at proper conclusions. In fact, counterexamples and edge cases are a frequent tool used to refine the understanding of a concept in a Socratic dialogue.

Reframing this in the context of artificial intelligence, and especially AI ethics, the Socratic example would translate into striving to find an adequate principle that can be used in a 'top-down' approach. However, this reinvigorated 'top-down' approach should not be seen as a complete substitute, but rather as a supplement to the 'bottom-up' approach. This is not a novel idea, however, as there are many scholars calling for the 'hybrid' approach (Kim et al., 2020). A 'Socratic hybrid' approach would entail formulating at least a tentative general rule of action, then providing a host of edge case examples showing exceptions or ways not to implement the general rule, monitoring the resulting output and then modifying the general rule accordingly. In a wish/command-giving context, this would mean issuing a command alongside examples of unwanted ways of its realization, prompting the AI agent to consolidate them and observing the output and adjusting the initial command until satisfactory results are achieved. Effectively, perverse instantiation is mitigated by teaching the AI what a perverse instantiation might be by the means of something similar to a Socratic dialogue.

Since the topic of perverse instantiation is closely tied to potential risks from superintelligences, we would be remiss not to reflect on how this legacy of Socrates can help in that area. A Socratic dialogue with a superintelligence aimed at aligning its goals with ours is, unfortunately, likely to be insufficient to mitigate the existential risk because having a dialogue by itself does not guarantee that a common standpoint will be achieved. An honest dialogue is predicated on the good faith of participants, and as the trial of Socrates shows, interlocutors of bad faith can have lethal consequences. A superintelligence 'acting in bad faith' is precisely what constitutes an existential risk from it. On the other hand, if we were to somehow manage to guarantee that a superintelligence will interact with us 'in good faith', that would mean that the problem of perverse instantiation is already solved.

Socrates would, at the very least, affirm the advice that proper precautions are essential for dealing with superintelligences. The underlying supposition of many discussions Socrates led with his contemporaries was that one should conduct thorough examinations of ideas before deciding to act on them. Evil, he thought, stems from ignorance. An erroneous idea can still be corrected, whereas a bad act cannot be undone. In a similar vein, knowing how to control a superintelligence must come before a superintelligence is given any agency. That truly is a great challenge, which is all the more reason to redouble our collective efforts in that direction.

It would seem that the best lesson we can take from Socrates is an organizational one. He was lauded as the wisest of the Greeks for knowing that he knows nothing, prompting him to seek out other knowledgeable people and engage them in polemical and productive conversation. In the contemporary context, this can be a model for a more immediate involvement of philosophers in AI development projects. After all, philosophers are notorious experts at conjuring up edge cases that bring established beliefs into question. Instead of philosophers orbiting around the AI field, offering general advice and opinions and being called up to comment only when specific trouble emerges, a more direct presence in AI projects may be beneficial – as in a dialogue, so to speak. For instance, by being intimately acquainted with a specific project, its scope and perspective, a philosopher might more effectively conceptualize possible perverse instantiations, whether great or small, that can then be taken into account before they potentially emerge in actuality. Such a role is ideally suited to educated philosophers who usually have good skills with critical and improvisational thinking. And the sooner we begin with this practice, the better off we all are, as each prior practical experience can better our relative position in the future.

But simply putting philosophers into AI development teams will definitely not be sufficient. A Socratic dialogue requires at least two participants, ideally more, and a singular and isolated philosophical standpoint is not enough. This brings us, at last, to the aspect of this conclusion that has the most far-reaching potential: AI developers can benefit greatly from education in humanities. This is not a new idea. Many educational institutions already have required humanities courses in their computer science programmes. Their value is usually explained as one of personal development: acquaintance with humanities is said to give a well-rounded education (Khalid et al., 2013) and to contextualize other knowledge (Bernard, 2022). But our analysis demonstrates that computer scientists can find content with direct significance for them in philosophy, history and literature in the shape of the cautionary stories we mentioned in this chapter. Each of those stories is a lesson learned in the past, and reviewing them in a contemporary

102   Goran Rujević

context is a way to make sure they stay learned. Furthermore, any objections that a humanities course is irrelevant for technical education can be intercepted by having a course that is tailor-made to highlight the connections like those we dealt with on the preceding pages.

We still need to address the most glaring issue of this proposal. Socrates only very rarely arrived at positive results of his inquiries. Would this mean that a Socratic approach would also tend towards dead ends? Furthermore, philosophical systems can sometimes seem as convoluted thought-mazes which are often in conflict with one another. Of the many systems of ethical norms, how can we know and choose the one that is true, if we can even reliably claim that there is one that can be called true? Intension and extension are parameters that are relevant for ensuring stable communication and in analysis of existing concepts or ideas. They, however, are not reliable tools for producing new ones, as their main function is critical.

The catch is that with regards to AI ethics, we do not need to chase after a single 'true' system. Owing to the possibility that artificial intelligences may end up being markedly different from human intelligence, we should regard the field of AI ethics as possibly independent from human ethics. That is to say, there is no guarantee upfront that ethical systems crafted with human agents in mind will be usefully applicable to AI agents. Instead of seeking the 'one true' universal and objective system of morality (if such a thing is at all possible), we should, at least in the very beginning, seek systems of norms that have the potential to be immediately useful and applicable and then interrogate these systems in search for inconsistencies, conflicting results or potentially troublesome interpretations. Of course, the most sensible beginnings should be from that which we already know, so in that regard, our initial approach may have a slight utilitarian slant.

The comparison undertaken in this article can be seen as one more reason in favour of hybrid approaches to teaching AI morality. Relying exclusively on top-down or bottom-up approaches invites potential shortcomings in the same sense in which using only intensional or only extensional definitions may lead to degenerate cases of misunderstanding. Furthermore, this article offers another point that affirms Yudkowsky's understanding that wishes are very complex things (Yudkowsky, 2011). Finally, the most important message is that AI ethics and safety ought to be regarded as a new area in which transdisciplinary approach, constant work, collaboration and refinement is not only to be expected, but is to be required. In a manner of speaking, by the use of our technology, we have added another dimension to the already rich conglomerate of eternal human projects.

This may sound disheartening at first, but an optimistic perspective is still possible. We can view this task as just another instance in which humanity ought to come together in our endeavours. Finding solutions for AI ethics and safety must be a team effort of people with diverse viewpoints, from programmers to philosophers, just like all Socratic discussion is between several people, young and old, amateurs and experts. No one person can know all things, no one person can see all the ways in which a command can be twisted into a perverse instantiation.

The only viable solution is to prepare ahead of time. Therefore, we ought to be careful and we ought to be humble.

References

Agar, N. (2016). Don't worry about superintelligence. *Journal of Evolution and Technology*, *26*(1), 73–82. https://doi.org/10.55613/jeet.v26i1.52

Aliman, N.-M., & Kester, L. (2019). Augmented utilitarianism for AGI safety. arXiv: 1904.01540 [cs.AI]. https://doi.org/10.48550/arXiv.1904.01540

Bernard, F. St (2022). Embracing humanities in computer science: An autoethnography. *Studies in Technology Enhanced Learning*, *2*(3). https://doi.org/10.21428/8c225f6e.d562588b

Bossman, J. (2016, October 21). *Top 9 ethical issues in artificial intelligence.* World Economic Forum. https://www.weforum.org/agenda/2016/10/top-10-ethical-issues-in-artificial-intelligence/

Bostrom, N. (2014). *Superintelligence; Paths, dangers, strategies.* Oxford University Press.

Plato. (1997). J. M., Cooper (Ed.), *Complete works.* Hackett Publishing Company.

Enfield, W. (1837). *The History of philosophy from the earliest periods: Drawn up from Bruckner's historia critica philosophiae.* Thomas Tegg and Son.

Frege, G. (1892). Über Sinn und Bedeutung. *Zeitschrift für Philosophie und philosophische Kritik*, *100*(1), 25–50.

Goodfellow, I. J., Shlens, J., & Szegedy, Ch (2014). Explaining and harnessing adversarial examples. arXiv:1412.6572 [stat.ML]. https://doi.org/10.48550/arXiv.1412.6572

Herodotus. (1904). *The histories of herodotus D.* Appleton and Company.

Jacobs, W. W. (1910). *The Monkey's paw.* Samuel French.

Khalid, A., Chin, C. A., Atiqullah, M. M., Sweigart, J. F., Stutzmann, B., & Zhou, W. (2013). Building a better engineer: The Importance of Humanities in Engineering Curriculum. *Paper presented at 2013 ASEE Annual Conference & Exposition, Atlanta, Georgia.* https://doi.org/10.18260/1-2-19270.

Kim, T. W., Hooker, J., & Donaldson, Th (2020). Taking Principles Seriously: A Hybrid Approach to Value Alignment. arXiv:2012.11705 [cs.AI]. https://doi.org/10.48550/arXiv.2012.11705

McGrath, Th., Kapishnikov, A., Tomašev, N., Pearce, A., Hassabis, D., Kim, B., Paquet, U., & Kramnik, V. (2021). *Acquisition of Chess Knowledge in AlphaZero.* arXiv:2111.09259 [cs.AI]. https://doi.org/10.48550/arXiv.2111.09259

Murphy, T. (2013). The First Level of Super Mario Bros. is Easy with Lexicographic Orderings and Time Travel... after that it gets a little tricky. In *A Record of the Proceedings of SIGBOVIK 2013* (pp. 112–133). Carnegie Mellon University. https://sigbovik.org/2013/proceedings.pdf

Pinker, S. (2018, February 14). We're told to fear robots. But why do we think they'll turn on us? *Popular Science.* https://www.popsci.com/robot-uprising-enlightenment-now/

Rawlins, G. J. E. (1997). *Slaves of the machine; The quickening of computer technology.* The MIT Press.

Shakespeare. (2003). *Macbeth.* Simon & Schuster.

Song, F., & Yeung, F. S. H. (2022). *A Pluralist hybrid model for moral AIs.* AI & Society. https://doi.org/10.1007/s00146-022-01601-0

Wallach, W., Allen, C., & Smit, I. (2008). Machine morality: Bottom-up and top-down approaches for modelling human moral faculties. *AI & Society, 22*(4), 565–582. https://doi.org/10.1007/s00146-007-0099-0

Weld, D., & Etzioni, O. (2009). The First Law of Robotics (A Call to Arms). In M. Barley, H. Mouratidis, A. Unruh, D. Spears, P. Scerri, & F. Massacci (Eds.), *Safety and Security in multiagent systems* (pp. 90–100). Springer. https://doi.org/10.1007/978-3-642-04879-1_7

Wittgenstein, L. (2009). *Philosophical investigations.* Wiley-Blackwell.

Yampolskiy, R. (2018). Predicting future AI failures from historic examples. *Foresight, 21*(1), 138–152. https://doi.org/10.1108/FS-04-2018-0034

Yudkowsky, E. (2004). *Machine intelligence research institute.* https://intelligence.org/files/CEV.pdfCoherent Extrapolated Volition.

Yudkowsky, E. (2011). Complex value systems in friendly AI. In J. Schmidhuber, K. R. Thórisson, & M. Looks (Eds.), *Artificial general intelligence. AGI 2011.* Springer. https://doi.org/10.1007/978-3-642-22887-2_48

Chapter 5

The Gradual Unavoidable Colonization of the Lifeworld by Technology

Kostas Theologou and Spyridon Stelios

National Technical University of Athens, Greece

Abstract

In this chapter, we introduce the concept of techno-constructionism that identifies with the Technical Construction of Reality (TCR) suggesting the replacement of mere constructivism or Social Construction of Reality. We discuss of course the idea of a theoretical investigation of the lifeworld or Lebenswelt through the lens of technology. Today, we are witnessing technology prevailing on most important functions and systems that were previously done in a completely different way and logic. Rapid advances in computer science have made engineering a dynamically expanding industry that has infiltrated all areas of social life. Artificial intelligence (AI)-based applications are not merely technical, but also cultural and ethical (regarding values). A new perception of reality emerges that ostracizes all kind of societal entities and human communities and alter even lifeworld itself, transforming it into a tech-lifeworld (a Technik-Lebenswelt). Against this perspective, it is important to identify and address the ethical challenges that the unavoidable evolution of technology will bring forward.

Keywords: Lifeworld; lebenswelt; systems (e.g. economy); state; technology; artificial intelligence; mechanization; ethics

Introduction

In this chapter, we introduce the concept of techno-constructionism that identifies with the Technical Construction of Reality (TCR) suggesting the replacement of mere constructivism or Social Construction of Reality. We discuss of course the idea of a theoretical investigation of the lifeworld or Lebenswelt through the lens of technology. Technology is taking over essential functions that were formerly

---

The Ethics Gap in the Engineering of the Future, 105–116

Copyright © 2025 Kostas Theologou and Spyridon Stelios

Published under exclusive licence by Emerald Publishing Limited

doi:10.1108/978-1-83797-635-520241006

performed in an entirely different manner and logic. Habermas' concept of the 'colonization of lifeworld', this time by digital technology and artificial intelligence (AI) applications, might manifest itself in today's societal setting. Because of the ubiquity of technology-based context, knowledge and information, the guiding means that control technology, particularly AI, are becoming more closely tied to human experience. An interaction between artificial and biological beings is emerging which defines the way Lebenswelt is being experienced and requires an ethical treatment in terms of its management and future consequences. In such a theoretical framework, we understand that Lebenswelt is being conquered by the former systems (e.g. economy, administration, state) which are now unified under their unique homogenization under AI and technology. A new perception of reality emerges to rapidly ostracize all kind of societal entities and human communities and alter even lifeworld itself. We could even claim a kind of eviction of human communities and societal entities by a both alien and familiar newcomer who intruded, inhabited and in now perceived as an innovator of the lifeworld, transforming it into a tech-lifeworld (a Technik-Lebenswelt). The entire planet and globalization itself may be thought and perceived through this novel concept of techno-constructivism that enhances us to better comprehend our tech-lifeworld.

Lebenswelt: the Conceptual Origins

According to Habermas, the colonization of the lifeworld is being established in modern capitalist societies as the instrumental and, rather, commodity logic of the capitalist system gradually dominates lifeworld relations, producing thus social pathologies. What is the notion of 'lifeworld' or 'Lebenswelt'? A starting point is the theory of Husserl (*Die Krisis der europäischen Wissenschaften und die transzendentale Phänomenologie. Ein Einleitung in die Phänomenologische Philosophie*, 1936/1970). Husserl's definition of the lifeworld is rather complicated; it may be thought of as the spectre of all our experiences, the background on which all objects appear as meaningful. This background is dynamic and not static, in the sense that nothing can appear except as lived. It is the objects variously arranged in space and time relative to perceiving subjects that constitute the ground for all shared human experience. It refers, among other interpretations, to a sense of meaning that can be established in cultural and social terms. The common lifeworld of a single community of objects could be perceived as the system of meanings that constitute common language, under the assumption that the subjects conceive of themselves and the world in the categories provided by that language. For subjects that belong to different communities, their shared lifeworld is a general framework of meanings which makes possible the mutual translation of their languages into one another. The concept of 'lifeworld' seems to denote how members of social groups (or group) structure their world into objects (Beyer, 2022).

Schütz (1945) adopted a more pragmatic approach claiming that lifeworld is not the private world of any of us; it refers to the world of immediate experience

The Gradual Unavoidable Colonization of the Lifeworld **107**

in which the taken-for granted takes place. It is our self-evident and taken-for-granted world, our fundamental and paramount reality. The world of daily life is dominated by eminently practical interests. It refers to the shared life on any person who lives amongst his peers. People operate in the 'natural attitude'; they take the world for granted and do not doubt its typifications and recipes until a problematic situation arises (Ritzer & Stepnisky, 2018, p. 595). Through a connection between all individuals, the lifeworld is recognizable and familiar forming our social world. In addition, because it is clear only to a limited extent, it contains contradictions (Vargas, 2020, p. 420).

Habermas sees social sciences and system theory as an extension of the ideas of Husserl and Schütz. Going further than the descriptive intentions of interpretative sociology (see Schütz), he emphasizes on the possibility of consensus as a guiding principle for sociological reflection. This seems to be in tension with a phenomenological account of social action. Language and discourse are the leading clues to the constitution of society (Pontin, 2022, pp. 1–2; Simpson & Ash, 2020).

Although Husserl was mostly concerned with the universal structures of experience, Habermas perceived Lebenswelt as a concept that may be used in a wider sense. Social and cultural structures of experience, common sense and language, as well as rules and laws that order common everyday activities are some of its components. Unlike Husserl, Habermas (1987) does not focus on consciousness. He examines the lifeworld through his theoretical tool of communicative action (*Theorie des kommunikativen Handelns*, 1981). The lifeworld consists of socially and culturally filtered linguistic meanings and is regarded as the environment of attitudes, practices and competences that are represented in terms of a subject's cognitive sphere. It represents the lived field of, mostly, culturally grounded informal understandings. In modernity, the lifeworld (people, society, culture, etc.) is disconnected from tradition (see the dominant institution of kinship) and founded on communicative action, realized broadly as a consensual form of social coordination for promoting common understanding in a group, cooperation and rational consensus.

Lifeworld can also be considered as a reformulation of the concept of social integration. Within modernity, it is grounded in communicative action and rationality. In the modern complex social structures, it is different, since the state, the market, the bureaucracy are now taking over the processing of important functions which in the past were being carried out in a completely different way (Fotopoulos, 2010, p. 337).

Colonization

For Habermas, there is a politico-economic dominance that has the effect of undermining the autonomy of other vital institutions within society. The market variable forces and the state have penetrated and 'colonized' the lifeworld, that is, the social and cultural space, preventing the development of important institutional parts of social reality. The hegemonic institutional sections of economy and politics are imposed in a way that undermines the autonomous functioning of the

108 Kostas Theologou and Spyridon Stelios

lifeworld. As the modern state becomes increasingly centralized and systematized, the communicative and consensual foundations of the lifeworld are, in turn, subjected to rationalization. For Habermas, this represents an 'inner colonization of the lifeworld', a destruction of the resources of cultural tradition as well as an erosion of the content of human relations (Elliott, 2022, p. 194).

The system's media of coordination (money, power, instrumental Reason) prevail over the lifeworld's means (family, religion, communicative reason). System imperatives and formally organized domains of action constantly occupy vital areas of social and cultural space and traditional rules cease to function as the main driver of social integration and cohesion. Colonization happens when critical disequilibria in material reproduction could be avoided only at the cost of pathologies or subjectively experienced crises of the lifeworld (Habermas, 1987, p. 305).

Habermas attempts, critically, to offer the possibility of emancipation against the impasse of modernity. Knowledge and rational discourse – realized through communication and language (see for instance, moral argumentation and legal-political discourse) could become tools of progress and democracy, not of domination and authoritarian rule. There is the potential to develop a new ethics of communicative action that will contribute to the progress of the social system. This possibility is possible even through the colonization of the lifeworld (Fotopoulos, 2010, p. 338). Society is evolving.

So, today's rationalization of everyday communicative practice is due to the penetration of administrative and economic rationality into areas of action specialized in social integration and cultural transmission; areas that are dependent on mutual understanding (Habermas, 1987, p. 330). The instrumental logic of the modern capitalist system dominates lifeworld relations and structures, creating social pathologies that arise when the communicative infrastructure of the lifeworld is 'colonized' by power and profit. Social integration and socialization that are interconnected with communicative action are encroached upon by money and power (see Habermas, 1986).

Communicative Action

The lifeworld is a concept complementary to the concept of communication action. Habermas (1987) presents three actor-world relations, clarifying how the lifeworld is related to the objective, the social and the subjective world. It is on those worlds that subjects, acting towards mutual understanding, base their common definitions of situations. Specifically, there are three different pragmatic relations that an actor/speaker can take up to something in a world: (a) to something in the objective world (see entities about which true statements are possible), (b) to something recognized as obligatory and shared by all members of a collective in the social world (see legitimately regulated interpersonal relations) and (c) to something that other actors (for instance, addressees) attribute to the speaker's own subjective world (see the speaker's unique experience which they can express before a public). So, the referent of the speaker's speech act is

The Gradual Unavoidable Colonization of the Lifeworld   *109*

presented to the speaker as something objective, normative (social) or subjective (Habermas, 1987, pp. 120–122).

Within a cooperative interpretation process, actors relate simultaneously to something in the objective, the social and the subjective world. This is done even when they thematically stress only one of the three. For example, when a hearer accepts the truth of an assertion (objective world), he/she acknowledges the other two implicitly raised validity claims such as the normative appropriateness of the specific utterance, or the sincerity of the speaker. Otherwise, the hearer is supposed to make known his/her dissent. For Habermas (1987, p. 121), this is a rule of communicative action. Actors are always expressing themselves in situations that they must define in common so far as they are acting with an orientation to mutual understanding.

To better understand the way in which these three relationships manifest and define everyday situations, let us examine an example presented by Habermas (1987, p. 121). In a construction site, an older construction worker who sends a newly arrived younger co-worker to fetch some beer, telling him to hurry up and be back in a few minutes, supposes that the situation is clear to everyone involved (all the workers that can hear the command). In this but also in similar cases, Habermas distinguishes a *theme, goals* and a *plan.*

> The theme is the upcoming midmorning snack; taking care of the drinks is a goal related to this theme; one of the older workers comes up with the plan to send the "new guy," who, given his status, cannot easily get around this request. (Habermas, 1987 , p. 121)

The normative (social) framework that allows the older worker to tell the younger one to do something is the informal group hierarchy of the workers on the construction site. In spatio-temporal terms, the action situation is defined by the distance between the construction site to the nearest store selling beer and by the upcoming break.

The lifeworld is the source of definitions of the *situation* (Fleming, 2002, p. 3). A *situation* is a segment of lifeworld contexts of relevance. It emerges by *themes* and is expressed through *goals* and plans of action. In this case, the situation from the lifeworld of the workers involved is marked off by the *theme* of the midmorning snack and the *plan* of fetching beer. The *situation* consists of elements which represent actual needs for mutual understanding as well as actual options. These include the status of a newly arrived younger co-worker, the workers' expectations concerning the midmorning snack, the distance of the store from the construction site and so on.

The background of this kind of communicative utterances is formed by situation definitions that involve correlating contents to worlds: what counts in each instance as an element of the objective world, as a normative component of the social world (see informal group hierarchy) or as an element of a subjective world. The theory of communicative action aims to offer a vision that allows the effects of colonization to come into perspective (Fleming, 2002, p. 4). The dominance of

110 Kostas Theologou and Spyridon Stelios

systems over the communicative infrastructure of everyday activities creates social pathologies such as alienation or reification, which, according to Lukács, broadly means mistaking social relations for things (Feenberg, 2015, p. 490).

Lebenswelt and Technology

It is not only the media of co-ordinations of the system, such as financial profit and power that prevail over the means of the lifeworld such as family or everyday life, but also technology in a mostly catalyst way. Traditional rules cease to function as the main locomotive process of social integration and cohesion. One could argue that by occupying vital spheres of human action, the technological colonization of the lifeworld is changing human behaviour. This hypothesis takes on greater value since the 21st century is often referred to as the digital age. Digital ways of communication and spreading of ideas are taking over, forming a dominant cyber culture. The lifeworld is being 'digitized' (Durt, 2022, p. 68). Our physical and the digital world are very constantly and increasingly interwoven. But they are not identical. If the digital and physical worlds were (ontologically) identical, then there would be no reason to investigate the ways in which one affects the other (Schneider, 2019). This difference between the lifeworld, as we know it, and its digital representation – serving as a complementary (?) element of lifeworld's manifestation – is also the main source of any ethical reflection. For example, speech acts, writing, thoughts and even perception are processes of our cultural self-understanding. The way they change is due to the pure and innate social upheavals of our daily routine and society and/or the modification of our behaviour due to technical-digital innovations structuring our environment (Schneider, 2019).

An example of the digitization of social behaviour is the various platforms that we use in our daily lives. These systems draw user data and provide customized recommendations to increase engagement. This is common in AI-powered applications. Through online search engines, AI provides more relevant results making it easier for users to find information and products. Within this framework, let us investigate now how Habermas' example of the older and newer worker could be reformulated within the modern digitized society. Let us imagine again a group of workers renovating an old house in an area they do not know. It is morning and they want to have their coffee. One of them, the oldest one, announces that it is time to order coffee and almost immediately searches his smart phone for the nearest shop, probably also looking at its service rating. The worker who gives the order assumes that the situation is clear to those involved.

When the younger worker (the hearer) assents to a thematized validity claim, that is, the online recommendation of coffee shop and its review score (normative/ social framework), he acknowledges the sincerity of the speaker or the normative appropriateness of his utterance. Consensus comes about, since the younger worker or the workers present accept(s) the truth of the online recommendation put forward by the older co-worker and at the same time does/do not doubt(s) the two abovementioned implicitly raised validity claims. As already mentioned,

The Gradual Unavoidable Colonization of the Lifeworld   *111*

participants acting with an orientation to mutual understanding express themselves in situations that they must define in common. The older construction worker who, in view of the morning coffee, immediately consults his mobile phone to make an online order, supposes that the situation is clear to everyone involved. He believes that with this action, all present, who will or will not have coffee, agree.

The *theme* here is the morning coffee at work; taking care of the coffee for all is a *goal* related to this everyday 'ritual'. The older worker comes up with the *plan* to 'google' the nearest and better, according to the reviews, coffee shop and make an order. All the other workers, younger on not, cannot easily get around this initiative, given the reliability of the internet as an information and service provider. The hearer(s) accept(s) the truth of that assertion (objective world). The specific *situation*, a segment of the workers' lifeworld context, emerges, amongst other things, by the morning coffee *theme* and it is expressed through the act of ordering online.

The delivery of the coffee in the working site removes the specific spatio-temporal dimension that is present in Habermas's original example. Nobody needs to know how far the store is, if a car is needed or if it is within walking distance. The suggestion of online ordering limits the chances of the hearers' engagement and denial of the speaker's claim. In the end, through this specific situation, it changes the lifeworld of the workers. Furthermore, through the possibilities offered by technology, the hierarchy in Habermas' example seems to be collapsing. Hierarchically superior actor in the group is now the search engine!

Since technology is being established in all social practices, we can now discuss about a new and rather objective (purely rationalized) technological construction of reality (TCR) or techno-constructionism rather than a vague and culturally based social construction of reality. We thus claim that in social construction of reality, there are several relativistic features (values, culture, economic development, social structure and formation, e.g. tribal, Muslim, Zealot, Amish, etc.) that turn a social construction of reality à-la-carte, while the TCR may foster a universal and cruelly objective perception of reality through the technological achievements and records; this perception is gradually turning participants (=the global population) to a kind of video gamers who adapt into the pseudo-reality of their remote controls and smart devices (technology-induced social pathologies).

In such a theoretical framework, we understand that Lebenswelt is being conquered by the former systems (e.g. economy, administration, state) which are now unified under their unique homogenization under technology. A new perception of reality emerges to rapidly ostracize all kind of societal entities and human communities and alter even lifeworld itself. We could even claim a kind of eviction of human communities and societal entities by a both alien and familiar newcomer who intruded, inhabited and in now perceived as an innovator of the lifeworld, transforming it into a tech-lifeworld (a Technik-Lebenswelt). The entire planet and globalization itself may be thought and perceived through this novel concept of techno-constructivism that enhances us to better comprehend our tech-lifeworld.

Intelligent Machines

The glittering chariot of technology moves at breakneck speed on the social racetrack and is driven by one strong and resilient horse called AI! In our everyday processes, we assign tasks to increasingly intelligent machines. By assigning these tasks to serve the purposes of our daily lives, we, in fact, grant them the privilege of shaping the way we act and think. When we rely, for example, on smartphone apps to remind us of friends and family's birthdays, we remove the mental obligation to remember important dates. In the age of streaming content, an online algorithm guides young and old by suggesting the next videos to watch. Through identification apps, we can recognize plants and foods simply by photographing them and at the same time we learn almost everything about them (see Samad et al., 2022).

Generative AI, a subset of AI involves using machine learning models to learn patterns and generate new unique outputs in the form of text or images. A typical example of this technology is writing assistants. They offer personalized AI guidance and text generation that allow human users to produce high quality content, instant draughts, ideas and replies. These apps are marketed as writing companions that understand what any user is trying to say, hence suggesting ways to make their writing more compelling, clear and even more authentic! Their creators assert that these apps are completely in tune with whatever someone is writing. Be that as it may, these are recommendation systems that are particularly useful in the age of digital communication (see Li et al., 2024).

In general, modern-day machines, artefacts and applications set the regulatory framework that boils down to a basic rule: 'If you want my services, then learn my operating rules and follow them.' Thus, there exists a 'mechanisation' of humans: modern technology enters our everyday life and mechanizes-systematizes our behaviour (Stelios, 2023, p. 221). Also, our everyday life has information overload. On our mobile phone, we carry a personalized encyclopaedia, our banking system, our work and a variety of other useful data. An ocean of information is in a device in our pocket, provided we have learnt the instructions for use!

'Mechanization,' at least as an effect on humans, is certainly nothing new. Its first appearance seems to come with the first tools. Before the Industrial Revolution (and even before), the pace of hand-tools and various inventions were linked to breathing rates and other rhythms of the human body (body rhythms: breathing, heartbeat, walking, etc.). The first hand-tools introduced new skills and new rhythms to work. Later, during the period of industrialization, inventions aimed at a deskilling of work and human body rhythms (Pacey, 1999, pp. 19 and 22). The machines move at their own independent pace. What has differentiated the 'mechanization' in recent decades is a shift in the purpose of digital, web-based, artefacts to searching and finding information. Information, which represents a cognitive content, seems to place 'mechanization', in principle, on a cognitive level. Machines 'revisit humans' through the latter's need and/or desire for information.

But can we talk about truly intelligent machines today? A starting point that may help investigate this question is the famous Turing test. One estimate is that

The Gradual Unavoidable Colonization of the Lifeworld   113

test, which attempts to show whether a machine can think, will be passed when the machine itself absorbs and uses all the data (e.g. through Big Data and the internet) necessary to make the impersonation of a human possible as well as successful. In modern society, we are most likely already observing the first traces of this artificial evolutionary path. AI-based programs are applied to 'big data', in the form of images, video and voice, serving various commercial and non-commercial purposes. Stochastic systems are functioning beyond the narrow deterministic model of algorithms. Computational intelligence absorbs vast amounts of data along with the 'noise' and errors that come with it, making the range of machine learning and provided information extremely large. For example, AI-controlled online malware creates more persuasive messages, evades detection more effectively and adapts better to each target. What it's essentially trying to achieve is disguise itself as a human user (Stelios, 2023, p. 219).

So, although the common belief is that humans will programme the machines to pass the test, the real success in the test will come when the machine has already programmed, especially through the supply and demand for information, the human in such a way that the latter accepts it as (in disguise) member of their species. This two-way process appears to have begun. Simple and complex machines are used and programmed by humans and, at the same time, they programme human behaviour (Stelios, 2023, p. 220).

Future, AI and the Ethical Factor

There exists the possibility, artificial applications will stop *mediating* only between human to human and human to the wider environment and start *interacting* with us as equal members. Obviously, today we do not yet have to remind, in various ways, Google's Gemini that it is just our assistant (that there is an informal group hierarchy) or share crude jokes about our relatives with Chat-GPT. Nevertheless, the digital/artificial world on the one hand and the natural/biological world on the other are in a continuous process of convergence. The closer they get to each other, the more machines will fit into our universe, moral or not. So, in the not-too-distant future, we could attribute responsibility, anger, envy or love to an AI-powered model. In fact, technology seems to be an aggravating factor for the deepening and cultivation of social withdrawal. The greater the familiarity with (potentially) artificially intelligent agents, the more the 'mechanization' process (Stelios, 2023, p. 227) is strengthened, resulting in technology occupying vital spheres of everyday human action.

AI is changing the physical arrangement of our social environment. It changes, also, our lifeworld. The lifeworld is not an assumed reality beyond experience. It is the world we experience, which is meaningful to us in everyday life. AI integrates into a 'digitized' lifeworld much more thoroughly than any other technology. It 'invades' the Lebenswelt by intelligently navigating and changing meaning and experience. It uses different, in comparison to human consciousness and understanding, means that often resemble human intelligence. According to Durt (2022, p. 68), what makes AI-based systems intelligent is not their apparent resemblance to human capacities, but the fact that they navigate and change the lifeworld in ways that make sense to humans. For instance, an AI-generated writing assistant must guess through the context what the user wants to say and suggests words accordingly, but this AI guesswork is very different from the multifactorial human estimation.

AI-powered systems use data. But data alone are not sufficient for experience and understanding. The specific intentional relation humans have to the world is missing. We literally experience the lifeworld and understand meaning, whereas computational AI does not literally do so – even if it appears to be doing that (Durt, 2022, p. 80). So, even though computational AI is gradually dominating the lifeworld, it can never be integrated in the lifeworld in the same way humans are. However, in functional terms, this difference in the perception does not seem to be of great importance. The hegemonic and guiding means that control technology still penetrate and 'colonize' the lifeworld. They penetrate the social and cultural space, changing or even stopping the development of important parts of social reality such as everyday interpersonal contacts.

If we now combine this penetration or invasion with the information-based 'mechanization' process, we could be more certain about the idea of a technological colonization of the lifeworld. In other words, the observed mechanization of humans is an indication of a colonization of the lifeworld not only by the media of co-ordinations of the system, such as money and (state) power, but also by high technology. This association of mechanization and colonization is especially important if examined through the lens of morality.

In fact, the greater the interaction with intelligent machines, the more plausible is the claim that these systems already possess an implicit level of moral functioning that becomes more apparent with time. The challenge is to identify these ethical parameters of human–machine interaction and manage them properly. How should we deal with 'intelligent' systems that coexist with us, how should we address them and take them into account in our decisions? These questions acquire a greater value if we assume that a (moral) machine will be able to choose, in the future, certain actions based on intuitive mechanisms, in the same way that humans do. The creation and learning of such an 'intuitive algorithm' are the mainstay of any attempt at achieving real moral conduct. Furthermore, it will enhance the interaction of these artificial beings with humans setting new standards of conduct.

Conclusion or the Unbearable Complexity of Lebenswelt

Buoyed by our technological achievements, humans only look towards the path of the future, indifferent to the possibility of gaps and missteps that may derail or even stop the evolution of our wise and all-powerful species. Can we avoid this possibility by addressing the moral dangers involved? The very nature of ethics narrows controllability. Morality, unlike most common intelligences, has a structure that is fluid; a labyrinthine composition that depends on spatio-temporal

The Gradual Unavoidable Colonization of the Lifeworld 115

factors, such as the time, culture and special circumstances. As are the cases with the Lebenswelt, it is quite complex.

These moral structural blocks become even more unstable at the artificial level. Artificial morality growing in the shadow of artificial intelligence, invisible but present, is a mind puzzle for humanity. A puzzle that concerns a complex coexistence of biological and artificial intelligent agents. This coexistence can become unbearable as humans are being 'mechanized' and the Lebenswelt is being colonized especially by AI-based technology. It does not only bring to the fore the hegemonic and guiding means that control technology but highlights the need for an ethical investigation that will prepare humans for future challenges.

In addition, the complexity of Lebenswelt after being colonized by technology and AI will be simplified. The lifeworld will be invaded and dominated by AI platforms and accessories producing thus a quasi-human but mainly a machine-human reality where construction of the broader view of the real cosmos will be driven by rationally processed social/human demands and needs. The social construction of reality (SCR) is gradually being absorbed by AI and makes part of a new and rather objective (purely rationalized) reality, that is, the TCR which will cause technology-induced social pathologies.

Colonization concerns all areas of engineering. One cannot ignore though that as a concept it clearly reminds us of space exploration. There are issues of colonization there too, but they only concern other planets. So, the parallel or implication of the human future colonization of other planets with the proposed technological colonization on Earth is quite interesting. In fact, it could be argued that the roots of colonization now extend beyond Earth. So, the technology of space exploration, colonization and exploitation of new planets offers fertile ground for new conceptualizations. These new or upgraded theoretical constructs could define, in moral terms, both the Lebenswelt of our species and perhaps that of any other alien intelligent species!

Finally, in this chapter, we introduced and discussed briefly, at a theoretical level, the technological colonization of the lifeworld and its transformation into tech-lifeworld. But is it technology itself that colonizes the lifeworld, or the state and economic system through technology. In other words, is technology and engineering just a tool? The answer is probably not. The technological phenomenon through the 'autonomy' in decisions and actions of future AI will be an omnipresent systemic agent not a means for the colonization of the lifeworld. As such it will take root in the communicative soil of the lifeworld changing its composition and creating new ethical challenges that must be identified and addressed.

References

Beyer, C. (2022). Edmund Husserl. In E. N. Zalta & U. Nodelman (Eds.), *The Stanford encyclopedia of philosophy*. Department of Philosophy, Stanford University. https://plato.stanford.edu/archives/win2022/entries/husserl/

Durt, C. (2022). Artificial Intelligence and its integration into the Lifeworld. In S. Voeneky, P. Kellmeyer, O. Mueller, & W. Burgard (eds), *The Cambridge handbook of responsible Artificial Intelligence: Interdisciplinary perspectives* (pp. 67–82). Cambridge University Press.

Elliott, A. (2022). *Contemporary social theory: An introduction* (3rd ed.). Routledge.

Feenberg, A. (2015). Lukács's theory of reification and contemporary social movements. *Rethinking Marxism, 27*(4), 490–507. https://doi.org/10.1080/08935696. 2015.1076968

Fleming, T. (2002). Habermas on civil society, lifeworld and system: Unearthing the social in Transformation Theory. *Teachers College Record on-line, 2002*, 1–17. https://mural.maynoothuniversity.ie/1058/1/HabermasTFleming.pdf

Fotopoulos, N. (2010). From one-dimensional man to the colonization of the Lifeworld: Herbert Marcuse and Jürgen Habermas against modernism. In S. M. Koniordos (Ed.), *Social thought and modernity* (pp. 325–346). Gutenberg. (in Greek).

Habermas, J. (1986). In P. Dews (Ed.), *Autonomy and solidarity: Interviews*. Verso.

Habermas, J. (1987). *The theory of communicative action, volume 2, Lifeworld and system: A critique of functionalist reason, translated by T. McCarthy*. Beacon Press.

Husserl, E., & Carr, D. (1936/1970). *The crisis of European sciences and transcendental Phenomenology*. trans. D. Carr. Northwestern University Press.

Li, Z., Liang, C., Peng, J., & Yin, M. (2024, May 11–16). The value, benefits, and concerns of generative AI-powered assistance in writing. *Proceedings of the 2024 CHI Conference on Human Factors in Computing Systems (CHI'24)*. https://doi. org/10.1145/3613904.3642625

Pacey, A. (1999). *Meaning in Technology*. MIT Press.

Pontin, F. (2022). Towards a phenomenological contribution for social criticism: A critique of the normative conceptions of the lifeworld in Habermas. *Civitas - Revista de Ciências Sociais, 22*, 1–10. https://doi.org/10.15448/1984-7289.2022.1. 41208

Ritzer, G., & Stepnisky, J. (2018). *Classical sociological theory* (7th ed.). SAGE.

Samad, S., Ahmed, F., Naher, S., Kabir, M. A., Das, A., Amin, S., & Islam, S. M. S. (2022). Smartphone apps for tracking food consumption and recommendations: Evaluating artificial intelligence-based functionalities, features and quality of current apps. *Intelligent Systems with Applications, 15*, 200103. https://doi.org/10.1016/ j.iswa.2022.200103

Schneider, D. (2019). Difference between algorithmic processing and the process of lifeworld (Lebenswelt). In D. Wittkower (Ed.), *Computer Ethics - Philosophical Enquiry (CEPE) Proceedings* (p. 18). International Society for Ethics and Information Technology. https://doi.org/10.25884/g90d-n566. https://digitalcommons. odu.edu/cepe_proceedings/vol2019/iss1/10

Schütz, A. (1945). On multiple realities. *Philosophy and Phenomenological Research, 5*, 533–576. https://doi.org/10.2307/2102818

Simpson, P., & Ash, J. (2020). Phenomenology and phenomenological geography. In A. Kobayashi (Ed.), *International encyclopedia of human geography* (2nd ed., pp. 79–84). Elsevier.

Stelios, S. (2023). Artificial intelligence or artificial morality? In J. Casas-Roma, J. Conesa & S. Caballé (Eds.) *Technology, users and uses: Ethics and human interaction through technology and AI* (pp. 204–232). Ethics Press.

Vargas, G. M. (2020). Alfred Schütz's life-world and intersubjectivity. *Open Journal of Social Sciences, 8*, 417–425. https://doi.org/10.4236/jss.2020.812033

Section 2

Biotechnology

Chapter 6

Ethical Aspects of Promises and Perils of Synthetic Biology

Ivica Kelam

Josip Juraj Strossmayer University of Osijek, Croatia

Abstract

Synthetic biology begins with the underlying assumption that life and life forms can be divided into parts and reassembled or redesigned according to the whims of their creators. Therefore, synthetic biology needs to be at the centre of ethical thinking since it engages the very concept of life and radically changes it. In this paper, we will investigate the phenomenon of synthetic biology through an ethical analysis of the unfulfilled promises and potential perils surrounding this technology. The paper consists of four parts. In the first part, we will deal with the problem of defining synthetic biology since it is a field in which many scientific disciplines meet and intertwine. The second part will present a brief history of systemic biology and the groundbreaking creation of Synthia, the first synthetic organism. The third part focuses on synthetic biology's potential benefits and some prominent ethical issues. In the fourth part, we will point out the problem of synthetic biology regulation. In conclusion, we will highlight the essential ethical remarks on synthetic biology and provide the impetus for further ethical debate.

Keywords: Synthetic biology; ethics; life; promises; perils; synthia

Introduction

Drew Endy, one of the pioneers of synthetic biology, set out the essence of synthetic biology: 'If you consider nature to be a machine, you can see that it is not perfect and that it can be revised and improved' (Van Est et al., 2007; p. 2). Craig Venter, the most famous practitioner of synthetic biology, in 2006 said of the potential of synthetic biology: 'This is the step everyone has always been

---

The Ethics Gap in the Engineering of the Future, 119–136
Copyright © 2025 Ivica Kelam
Published under exclusive licence by Emerald Publishing Limited
doi:10.1108/978-1-83797-635-520241007

120   Ivica Kelam

talking about. Once we have learned how to read the genome, we can now also start writing it' (De Vriend, 2006, p. 9). In short, synthetic biology, as a completely new scientific field, strives for complete control of the basic building blocks of life. According to synthetic biologists, control is achieved through the design of organisms to have useful functions for society, for example, in health, energy and environmental protection, as we will see in the paper. Synthetic biology is a relatively young scientific discipline that has emerged in the last 20 years, on the border between molecular biology, biotechnology, organic chemistry, engineering and informatics/systems biology. Due to the efforts to radically transform life and life forms, synthetic biology should be at the centre of ethical considerations.

The paper consists of four parts. The first part will define itself as the vast synthetic biology field in which to meet and intertwine many scientific disciplines. The second part will present a brief history of systemic biology and the groundbreaking creation of Synthia, the first synthetic organism. The third part focuses on synthetic biology's potential benefits and risks. We will explore the benefits by analysing the possibilities of efficient pharmaceutical manufacturing. Issues of biosafety and biosecurity are the most vivid risks of synthetic biology. In the fourth part, we will point out the problem of synthetic biology regulation. In conclusion, we will highlight the essential bioethical remarks on synthetic biology and provide the impetus for further ethical debate.

What Is Synthetic Biology – In Search of a Definition

Defining what 'synthetic biology' is and what processes it encompasses is still complicated today, as there is not just one formal definition that includes all dimensions of synthetic biology. For instance, synthetic biology is, according to Voosen (2013), 'arguably the world's hottest and most poorly defined scientific discipline'. Schmidt confirms this statement: 'If you look for 10 experts to define synthetic biology, you will probably get ten different answers' (Schmidt, 2010, p. A118). Nevertheless, it is not just the problem of experts and their definitions of synthetic biology; various institutions also have their interpretations. Below, we will list some definitions and try to find the most accurate description of synthetic biology. In a 2009 report on synthetic biology, the Royal Academy of Engineering defines: 'synthetic biology is an emerging area of research that can broadly be described as the design and construction of novel artificial biological pathways, organisms or devices, or the redesign of existing natural biological systems' (Royal Academy of Engineering, 2009). According to 2009 the President's Commission on the Study of Bioethical Issues report: 'synthetic biology is a scientific discipline that relies on chemically synthesised DNA, along with standardised and automatable processes, to address human needs by the creation of organisms with a novel or enhanced characteristics or traits' (President's Commission on the Study of Bioethical Issues, 2010, pp. 46–47). The European Commission Scientific Committees, 2014 report definition: 'SynBio is the application of science, technology, and engineering to facilitate and accelerate the design, manufacture and/or modi-fication of genetic materials in living organisms' (European Commission Scientific Committees, 2014). The National Bioeconomy Blueprint 2012 report defines synthetic biology: 'the design and wholesale construction of new biological parts and systems, and the redesign of existing, natural biological systems for tailored purposes, integrates engineering and computer-assisted design approaches with biological research' (National Bioeconomy Blueprint, 2012; p. 15). In 2016, at the 13th Conference of the Parties (COP) to the Convention on Biological Diversity held in Paris, the following definition of synthetic biology arose: 'a further development and new dimension of modern biotechnology that combines science, technology and engineering to facilitate and accelerate the understanding, design, redesign, manufacture and/or modi-fication of genetic materials, living organisms and biological systems' (Convention on Biological Diversity, 2016).

All the definitions, as mentioned earlier, have fundamental agreement on synthetic biology as being 'the deliberate design and construction of novel bio-logical parts and systems for pre-identified purposes'. Synthetic biology is a field of research whose main objective is to create fully operational biological systems from the most minor constituent parts possible, including DNA, proteins and other organic molecules. Synthetic biology incorporates many scientific tech-niques and approaches, which we will explain below (Rugnetta, 2024).

A Brief History of Synthetic Biology

The 'term synthetic biology' appeared first time by the French scientist Stephan Leduc in 1910, in the book *Théorie physico-chimique de la vie et générations spontanées*, which contained a chapter entitled *La Biologie Synthetique* (Leduc, 1910). In 1912, Leduc wrote La Biologie Synthetique (Leduc, 1912). The phrase synthetic biology was used again 62 years later in a 1974 paper by geneticist Wacław Szybalski, who wrote:

> Let me now comment on the question 'what next'. Up to now, we are working on the descriptive phase of molecular biology.. But the real challenge will start when we enter the synthetic biology phase of research in our field. We will then devise new control elements and add these new modules to the existing genomes or build up wholly new genomes. This would be a field with the unlimited expansion potential and hardly any limitations to building 'new better control circuits' and..... finally other 'synthetic' organisms, like a 'new better mouse'.. I am not concerned that we will run out of exciting and novel ideas,... in synthetic biology, in general. (Szybalski, 1974, p. 405)

Four years later, Szybalski and Skalka, in an editorial of the journal *Gene* from 1978, written on the occasion of the award of the Nobel Prize for the work of Werner Arber, Daniel Nathans and Hamilton Smith on restriction enzymes, enthusiastically point out that:

122   Ivica Kelam

> The work on restriction nucleases not only permits us easily to
> construct recombinant DNA molecules and to analyse individual
> genes but also has led us into the new era of 'synthetic biology'
> where not only existing genes are described and analysed but also
> new gene arrangements can be constructed and evaluated.
> (Szybalski & Skalka, 1978, pp. 181–182)

After the editorial mentioned above, the title 'synthetic biology' again appeared in 1980, when Barbara Hobom used it to describe bacteria that had been genetically engineered using recombinant DNA technology (Hobom, 1980, p. 14). Because these bacteria are living systems (hence biological), they are altered by human intervention (i.e. synthetic). We can conclude that, according to Hobom, synthetic biology is mainly synonymous with 'bioengineering'. In 2000, at the annual meeting of the American Chemical Society in San Francisco, Eric Kool stated, 'I'm calling this approach synthetic biology' (Rawls, 2000, p. 49) and explaining the synthetic capability of organic and biological chemistry to design non-natural, synthetic molecules that nevertheless function in biological systems. He was among the first to use synthetic biology in a rather 'contemporary' understanding. From an academic standpoint, the critical event for the development of synthetic biology was the conference 'Synthetic Biology 1.0' in June 2004, held at the Massachusetts Institute of Technology. Synthetic Biology 1.0 was the first international conference dedicated to synthetic biology research (Ball, 2004). The conference attracted much attention in the scientific world due to its interdisciplinarity. It gathered many professionals ranging from biology and chemistry to computer science, and the central emphasis was on designing, building and characterising biological systems and interactions (Ferber, 2004). The conference in 2004 was the first in a series, and the last SB7.0 was held in Singapore in June 2017. Cameron et al. point out that this series of conferences sparked a debate about combining engineering elements with molecular biology, with the general goal of describing whether synthetic biology could become as developed an engineering field like electrical engineering or materials science (Cameron et al., 2014). The period from 2004 to 2010 was called 'the second wave of synthetic biology', as various elements of synthetic biology began to take shape in circuit design and metabolic engineering (Purnick & Weiss, 2009). Finally, a significant breakthrough was achieved in 2010 when the first organism with a synthetic genome was created, as will be described below.

Creating Artificial Life From Scratch – Minimal Genome Project

In 2010, the general and scientific public was stunned by a paper published in *Science* that a team of scientists led by synthetic biology pioneer Craig Venter had created the entire genome of a bacterium called *Mycoplasma mycoides* in his lab and implanted DNA into an empty cell of another related microbe (Gibson et al., 2010). They called it Synthia. Media headlines screamed from the front pages that artificial life was created for the first time (Sample, 2010). Those familiar with the issue noted that they were copying life, putting the existing genome into a new chassis, like 'hermit crabs residing in an abandoned shell' (Lewis, 2013). This project, called a minimal genome project, aimed to identify the minimal set of genes needed to sustain life from the genome of *M. mycoides* to regenerate those genes synthetically to create a 'new' organism. Building the first 'synthetic' organism is a time-consuming and expensive process, confirmed by estimates that it took 15 years of research from initial idea to final realisation and over $ 40 million in cost (Pennisi, 2010, p. 958). However, it is important to emphasise that no matter how impressive it looks, no synthetic life was created by creating Synthia. Life would not be possible if over a one million base pair *M. mycoides* genome was not synthesised and implanted into a DNA-free bacterial shell of *Mycoplasma capricolum*. Then *M. capricolum*, due to its self-replication, enabled the emergence of a new 'synthetic life'. Despite this, a team of researchers from the Venter Institute continued their work, and in 2016, they announced that they had managed to synthesise another even smaller genome, which contains 531,560 base pairs and 473 genes. A comparison of these two genomes revealed a standard set of 256 genes that the team thinks could represent a minimal set of genes needed for viability (Hutchison et al., 2016). The critical problem is that out of 473 genes, as many as 149 of these genes are of unknown function. In the words of Craig Venter:

> We don't know about a third of essential life, and we're trying to sort that out now. We've sequenced everything on this planet, and we still don't know 149 genes that are most essential for life! This is the coolest thing I want to know. (Callaway, 2016, p. 558)

The Veneter statement perfectly depicts all the complexity of creating a 'new synthetic' life. Although this is a major scientific breakthrough, it once again confirms that life is still incredibly complex and unknown, even in such simple life forms as the bacteria, as mentioned above. This project shows us that creating synthetic artificial life is not an easy task and, from the standpoint of bioethics, is a very questionable justification for such ventures that involve the very essence of life. Bearing in mind that in this particular case, for as much as a third of genes, it is not known at all what their function is in the genome, except that they are crucial for the origin and maintenance of life.

Potential Benefits and Ethical Issues of Synthetic Biology

More Efficient Manufacturing of Pharmaceuticals

According to synthetic biologists, one of the benefits of the application of synthetic biology is the faster and more efficient production of vital materials for specific drugs and vaccines (Paddon & Keasling, 2014). One of the first examples is Jay Kiesling, a professor at Berkeley, one of the leading experts in the field of synthetic biology, the development to produce semi-synthetic artificial artemisinin for treating malaria patients. Kiesling's process included synthesising antimalarial precursors in artemisinic acid, providing a faster, cheaper and more reliable alternative to traditional artemisinin used to alleviate malarial conditions in patients (Singh & Vaidya, 2015). Sanofi has, by 2016, produced more than 39 million semi-synthetic artemisinin treatments, representing about 10% of the demand for global artemisinin-based combination therapies (Peplow, 2016). Another early attempt was to mass-produce the influenza vaccine in a shorter time frame than within conventional production measures. Namely, Novartis began in 2012 an effort to use synthetic biology to shorten the time required to produce an influenza vaccine (Rojahn, 2013). According to Dormitzer et al., the first case of this technological process included influenza strain H7N9. The entire process, from gene sequencing, synthesising, modifying and engineering virus material on a computer to growing material needed for early vaccine production, was approximately 100 hours. The vaccine production time was shortened from 6 months to just one week (Dormitzer et al., 2013). In addition to the above application, it is working intensively on developing engineered probiotics to aid digestion and improve overall human health (Bugaj & Schaffer, 2012). Primarily refers to the production of synthetic probiotic bacteria, designed to exhorted health benefits to users, such as improvement in the body's immune system, reduced gastrointestinal discomfort, decreased potentially pathogenic microorganisms, improved bowel regularity and others (Danino et al., 2015). Also, Weber and Fusnegger point out that the impact and potential of synthetic biology for biomedical applications can significantly improve health benefits in preventive and acute medical care.

On the other hand, health benefits generated from such novel probiotics, pharmaceuticals and vaccines pose a potential risk due to uncertainty from the in vivo use of synthetic biological agents (Ruder et al., 2011). Similar risks warn Church et al. when they call for extra caution due to the potential for unique health risks of such pharmaceuticals, with horizontal gene transfer (HGT) as a new risk (Church et al., 2014). However, the possibility of HGT within human and environmental cells is minimal, as shown below. Nevertheless, Getino et al. call for further research to help us mitigate the horizontal transfer of synthetic genetic material (Getino et al., 2015). Scientists fear that such probiotics can cause acute and risky side effects to an individual's gastrointestinal systems, such as the disruption of 'good' bacteria populations and the uncontrolled production of unhealthy gut bacteria. However, there is no scientific evidence for this deficiency. Still, such findings remain uncertain due to the lack of clinical testing (Tang et al., 2013).

Energy Production – Biofuels

In the early 2000s, with the initial development of synthetic biology, one of the most significant potential applications of this technology was considered to be the production of algae biofuels. According to some estimates, between 2005 and 2012, hundreds of millions of dollars were invested in algae-derived biofuels. For example, in 2007, the oil company BP invested $ 500 million in a ten-year

Ethical Aspects of Promises and Perils **125**

partnership with the University of Berkeley to research new energy sources and reduce the impact of energy consumption on the environment (Washburn, 2007). Two years later, in 2009, the largest oil company, ExxonMobil, announced a $ 600 million investment in the research and development of algae biofuels as part of a partnership with biotechnology company Synthetic Genomics, founded and owned by Craig Venter (Howell, 2009). Companies sprung up around technology to extract oil from these simple seaweeds. Under the right conditions, algae can produce high quantities of lipids. The key was to find high-lipid-producing strains and then, with the help of synthetic biology, expand the production of those lipids and turn them into fuel at a low enough cost to out-compete the actual price of oil. Because of all this, unrealistic promises have been made about a green bio-economy that will replace the fossil fuel economy with an environmentally friendly fuel that will benefit the environment and, at the same time, help combat climate change by removing CO_2 from the atmosphere. According to common belief, algae only need water and CO_2 to grow. After the initial enthusiasm, the reality showed that the expectations were completely unrealistic. In 2012, the US National Research Council released a report concluding that: 'the scale-up of algal biofuel production sufficient to meet at least five per cent of US demand for transportation fuels would place unsustainable demands on energy, water, and nutrients with current technologies and knowledge' (National Research Council, 2012, p. 2). A couple of years later, the European Energy Algae (Enalgae) project, which ran from 2011 to 2015, come to a similar conclusion:

> ...one of the major ideas enthusiastically considered five years ago was the potential role of microalgae in energy generation. With the barrel cost of oil almost halving and revised estimates for the realistic potential for algal biofuels coming from the Enalgae project, it now looks improbable that microalgae can contribute significantly to Europe's need for sustainable energy. (EnAlgae Consortium, 2015)

And finally, in 2017, a significant report from the International Energy Agency (IEA) on algae biofuel technology states: 'microalgae-based production to produce bioenergy products like liquid or gaseous fuels as primary products is not foreseen to be economically viable in the near to intermediate term' (ieabioenergy.com., 2017). These reports show that algae biofuels have no economic justification despite the enormous time and money invested in their development.

Nevertheless, promoters continue to promise that with the help of synthetic biology, algae will produce sufficient amounts of biofuels someday. However, proponents forget that algae are enormously diverse, and it is estimated that there are about 800,000 species, of which only perhaps 50,000 have even been described. Algae species thrive in oceans, freshwater, soils and tree barks, as symbionts with various animals (for example, in sloths' fur), and in many other environments. We do not know anything about most species; even for familiar species, we know little about their natural history or behaviour (biofuelwatch.org., 2017). Due to the enormous ignorance, there are numerous risks from the production of algae bio-fuels, and here we will mention the most important. The first problem is escaping containment since many scientists rightly assume that algae under cultivation in open ponds or photobioreactors cannot be prevented from escaping into the environment. Snow and Smith, in their paper, ask the following questions: 'How frequently would GE microalgae escape from cultivation and processing facilities? Escape into the environment could occur through aerosolisation, wildlife vectors, turbulent weather that damages or destroys these facilities, accidents, human error, or other events. How far would GE microalgae disperse, and how long would they survive? Could transgenes designed to enhance the growth and fitness of released GE microalgae subsequently spread across meta-populations, species, habitats, and regions, and, if so, at what scales and over what time frames?'

Other authors consider similar scenarios of the ecological impact of algal escape and the impossibility of preventing their impact. They conclude that the spread of GM microalgae of the configuration we identify would be impossible to halt. As GM factors are likely affecting the palatability of microalgae and are already being conducted in the name of biofuel production, there is a real risk that the genie is already partway out of the bottle. Suppose GM biofuels-optimised microalgae were to destroy fisheries. In that case, a primary driver for microalgae biofuels research, the argument that such biofuels would not compete with biomass production for food, maybe totally misplaced. Accordingly, a strong argument can be made for the regulation of GM microalgae at an international level because the potential for damage could have global consequences, echoing recent concerns over geoengineering. Whether against arguments for sovereign fuel security, regulation could be enforced is a dilemma that society may soon face (Flynn et al., 2012). Finally, a potential problem is HGT, as algae frequently engage in HGT. According to Beacham et al:

> ...a major concern for GM microalgae use is, therefore, that the modifications created may be transferred from the GMO via HGT into natural algae, bacteria or virus species in the environment and thereby cause damage to ecosystems via selective advantage conferred by the transferred genes. (Beacham et al., 2017, p. 96)

We can conclude that given their fundamental role in earth systems, it seems especially dangerous to manipulate and engineer algae for commercial and industrial uses, given the fact that 'containment' in production facilities is virtu-ally impossible. Since algae are fundamental to ecosystems and regulate ocean biogeochemical cycles, there is potential for severe and far-reaching harm. Besides, algae are the source of much oxygen that makes the earth habitable for humans and most other species, meaning their presence and population dynamics have broad-ranging consequences. They also form the base of the aquatic food chain; hence, the composition of algal communities is a defining feature of eco-systems. Due to the absence of knowledge about most algae species' basic biology and genetics, it is impossible to fully predict or control the impacts associated with

Ethical Aspects of Promises and Perils **127**

introducing modified algae into natural ecosystems (biofuelwatch.org. 2017). In the ethical assessment of this issue, considering all the listed potential environmental problems, the precautionary principle should be adhered to, and the cultivation of modified algae should be avoided.

Ethical Issues of Synthetic Biology

Although synthetic biology can potentially significantly improve human progress, for example, in reaching new scientific knowledge and its applications in areas from medicine to food and energy production, in the following, we will briefly analyse the potentially most significant ethical issues regarding the potential dangers of synthetic biology with particular emphasis on the concept of biosecurity and biosafety (Kelle, 2012).

Issue of Biosafety

Biosafety is a challenge to managing and administering any undertaking engaged in producing material or technology production, with specific safeguards for material handling, packaging, transportation and safe use required by statutory law and regulatory practice. Research with biological organisms has specific biosafety requirements based on the type of organism, its toxicity/pathogenicity/other concerns and whether the organism is considered native or invasive to the local environment (Burnett et al., 2009). What makes synthetic biology different from other technologies is the incomplete and insufficient understanding of the types of phenotypic traits and possible hazard scenarios that a constructed organism can possess, as well as how it will engineer organisms behaving and surviving in a different environment from that for which it was designed. Wright et al. point out that the most important thing is to prevent the unintentional genetic release that may subsequently alter or overwhelm its local environment and incur negative health consequences in considering the issue of biosecurity (Wright et al., 2013). One potential biosafety risk concerns the inadvertent release of genetic material that has been described as potentially problematic for several reasons, including (i) the potential for engineered organisms to act like invasive species and negatively impact biodiversity, (ii) concerns of exposure of engineered organisms to unintended human and animal targets and (iii) the inability to control engineered organisms – particularly bacteria – once they are taken outside of a contained environment. Synthetic biology adds another dimension of uncertainty and risk in biosafety concerns due to the potential release of substantially genetically modified organisms into the environment in an irreversible manner (Trump, 2020). For this reason, researchers suggest various constraint options to limit or prevent such biosafety events, including (a) engineered control options like genetic kill switches or reliance upon particular food or nutrient sources that are not easily found in the environment and (b) traditional safety checks such as oversight committees, proper biological containment protocol and safety gear, and approval requirements to edit or engineer certain genetic strains (Wright et al., 2013, p. 1230). Here, we will mention that due to

128   Ivica Kelam

biosafety concerns, the International Genetically Engineered Machine, or iGEM, is an example of a nonstate actor with rigorous biosafety requirements. IGEM is the most prestigious international competition that encourages high school and college students to research synthetic biology through mentored teams with strict safety and judging criteria (Guan et al., 2013).

Another potential biosafety risk concern is the concept of HGT. HGT is 'the non-genealogical transmission of genetic material from one organism to another' (Goldenfeld & Woese, 2007, p. 369). Generally referring to the transfer of genes between organisms in a manner other than traditional reproduction, HGT is a particular problem of concern for synthetic biology as such gene transfer is a common and somewhat uncontrolled trait through the microbial biosphere (Dröge et al., 1998). The experts discovered that HGT occurs by transduction, conjugation and transforming modified cells within the natural environment. For these three transfer methods, transduction involves the active transfer through bacteriophages, conjugation through pili and transformation via sequence-independent uptake of free DNA from the environment (Davison, 1999, p. 78). Although HGT is extremely rare, it can still occur, and the consequences can be irreversible and potentially extremely harmful to the environment.

Consequently, experts in synthetic biology have begun to explore avenues to prevent HGT via one or more of these avenues. The problem is the possibility of remaining modified cell DNA in the environment after the cell's death (Thomas & Nielsen, 2005). Although scientists point out that transferred traits are rarely evolutionarily beneficial to targeted organisms, they also harm human and ecosystem health in a natural setting (Rossi et al., 2014). However, White and Vemulpad note that synthetic biology may increase the potential for harmful gene transfer due to the use of artificial gene sequences. Therefore, they call for precautionary measures such as 'the use of more fastidious hosts and non-transmissible vectors for the synthetic genes' (White & Vemulpad, 2015, p. 62). Moreover, even if such a scenario materialised, Armstrong et al. argue that such concerns are more likely in deliberate biosecurity situations rather than through accidental release and random gene transfer (Armstrong et al., 2012).

In conclusion, we should be cautiously optimistic regarding HGT. In Werner's words:

> We can predict that future investigations will continue to offer welcome availabilities of identified genes and their products which can serve for beneficial use to the service of humankind and its environment. Likely, horizontal gene transfer will thereby continue to play a significant role. (Arber, 2014, p. 221)

Issue of Biosecurity

Biosecurity, or concerns of risk driven by the use of synthetic biology for nefarious or deliberately harmful means (i.e. bioterror), is the biggest fear in the public. Moreover, Marris calls this public fear synbiophobia and emphasises it as the

Ethical Aspects of Promises and Perils   *129*

driving force behind promoting public engagement and other activities in the field of synthetic biology (Marris, 2015). Many authors point out that due to the dilemma of dual-use, synthetic biology also poses a significant threat to humanity (Atlas & Dando, 2006). Tucker and Zilinskas point out that the problem is that any new invention and technique in synthetic biology can be used for nefarious purposes, such as the development of bioweapons – a matter of biosecurity or bioterror (Tucker & Zilinskas, 2006). Synthetic biology has faced a dual-use problem from the very beginning, as evidenced by reports jointly published by the J. Craig Venter Institute, the Center for Strategic and International Studies, and MIT:

> Synthetic genomics… is a quintessential 'dual-use' technology – a technology with broad and varied beneficial applications, but one that could also be turned to nefarious, destructive use. Such technologies have existed since humans first picked up rocks or sharpened sticks. However, biology brings some unique dimensions: given the self-propagating nature of biological organisms and the relative accessibility of powerful biotechnologies, the means to produce a 'worst case' are more readily attainable than for many other technologies. (Garfinkel et al., 2007, p. 1)

It should be noted that synthetic biology can create three aspects of the dual-use problem. First, the engineering of organisms will soon be more comfortable and cheaper, consequently making it easier and less expensive to produce pathogenic and potentially harmful substances by metabolic engineering. The rise of the DIY-bio movement raises concerns that synthetic biology is becoming available to people with less knowledge and is less reliable and transparent than university scientists and synbio companies. Therefore, the possibility of fatal accidents or the deliberate creation of pathogens cannot be ruled out (Kelle, 2013). Second, better knowledge in designing organisms increases the probability of 'designer pathogens' being intentionally made weaponised and thus more dangerous than similar natural pathogens. Because of this specific risk, synthetic biology has ended up on a list of '12 risks to human civilisation' prepared by Future of Humanity Institute researchers (Pamlin & Armstrong, 2015). The third aspect refers to publishing the genomes of known pathogens like smallpox or H5N1 virus in the public domain (i.e. scientific journals). So evildoers can get a blueprint online to make a devastating weapon. In bioethics, the third aspect is one of the most pronounced synthetic biology risk issues. A bioethical discussion has developed around this issue. For example, Douglas and Savulecsu, in their paper, call for the development of an 'ethics of knowledge' that will guide us in deciding whether to publish information that significantly increases the ability of evil people to produce dangerous pathogens (Douglas & Savulescu, 2010, p. 687). Some scholars like Michael Selgelid propose a solution to the stricter regime of government regulation and the prohibition of self-regulation (Selgelid, 2007).

Synthetic Biology Regulation

The discussion about how to regulate synthetic biology is always related to possible risks to human health and the environment. Therefore, the main question is whether synthetic biology poses unprecedented risks and requires specific regulation or can be covered by existing regulations of similar products. A global coalition of over 100 environmental, social, scientific, indigenous and human rights groups – including several from Africa, Asia, and Latin America – jointly endorsed a framework they said should guide the research and commercialisation of synthetic biology technologies. They presented a document called Principles for the Supervision of Synthetic Biology (FOE, ICTA, ETC Group, 2012). Eric Hoffman, head of the genetic technology campaign at Friends of the Earth in the United States, told the media it was 'the first civil society document to describe how synthetic biology should be regulated' (Sharma, 2012). The document advocates the application of the precautionary principle, which would mean a moratorium on the release or commercial use of synthetic organisms, cells and genomes until regulatory bodies have considered the risks – and a total ban on any attempt to change the human genome, stating: 'Any alterations to the human genome through synthetic biology – particularly inheritable genetic changes – are too risky and fraught with ethical concerns' (FOE, ICTA, ETC Group, 2012, p. 5). The document also called for the urgent adoption of mandatory regulations relating to synthetic biology aimed at protecting public health and safety of workers and the environment and for any regulation framework to include requirements for corporate accountability and manufacturer liability. As in the case of genetic modification, key stakeholders in the synthetic biology industry support the principle of self-regulation, as any regulation by the state represents an unnecessary barrier to the development and rapid introduction of new products to the market.

For example, Maurer, Zoloth and Bügl et al. advocate regulation governed by synthetic biologists instead of government regulations and thus want to prevent the introduction of national regulation (Bügl et al., 2007). Maurer and Zoloth emphasise the need for self-management without external intervention or intrusive supervision. Bügl et al. advocated a governance structure that would formally include external government oversight, but in essence, synbio companies would be in charge of regulation (Maurer & Zoloth, 2020). In a similar tone, Grunwald describes the former as generally precautionary in approach, which, while reducing the probabilities of bioterrorism and the same risks, would potentially limit progress by reducing the ability to disseminate knowledge freely and quickly (Grunwald, 2012). Here, we will briefly explain the fundamental difference in the regulation of synthetic biology between the USA and the European Union (EU). Considering differences in regulation between the USA and the EU, the fundamental dilemma is whether to regulate the product or the process. The discussion of a process/product related to synthetic biology is reminiscent of a similar discussion related to traditional genetic engineering. The EU has established the process regulation of GMOs, and synthetic biology is regulated by these regulations, namely Directive (EU) 2018/350. EU Directive 2018/350 requires that all

Ethical Aspects of Promises and Perils   **131**

products resulting from the process of synthetic biology, which at any stage include this Directive, regulate genetic modification in any form (Official Journal of the European Union, 2018). Quite the opposite is the case for product-based regulation in the USA. The products of genetic engineering, including synthetic biology, are regulated by the type of products. Therefore, synthesis-based drugs are regulated by the Federal Food, Drug, and Cosmetics Act (FDCA) of the Food and Drug Administration. Synthetic organisms used as pesticides fall under the Federal Law on Insecticides, Fungicides, and Rodenticides (FIFRA) of the Environmental Protection Agency. The Toxic Substances Control Act covers industrial use, such as the production of chemicals using metabolic engineering, and is also administered by the EPA (Carter et al., 2014).

In conclusion, the ethical assessment of regulation is a fundamental issue for this topic. Finding the right balance between advances in science and precautions is not easy. Therefore, the EU's position in regulating the process gives scientists enough freedom. At the same time, the precautionary principle provides us with some certainty that there will be no irreversible harmful consequences for human health or the environment.

Conclusion

In the ethical assessment of synthetic biology, it is necessary to transcend sensationalist headlines from newspapers with potentially catastrophic future scenarios in which synthetic Armageddon threatens to wipe out humanity. Nevertheless, we should also not be carried away by the euphoric promises of promoters who see synthetic biology as a panacea and solution to all problems. Climate change, species extinction and energy transition to faster and easier production of new drugs and vaccines are just a few recent examples. The fear and distrust of the public and various environmental, social, scientific and human rights groups towards the promises of synthetic biology are understandable. Given that a similar narrative has already taken place recently, here we are primarily referring to the case of genetically modified crops. In the preface to my book Genetically Modified Crops as a Bioethical Issue, I wrote that genetically modified crops are a means of power through which private biotech corporations privatise life forms, thus becoming owners of life (Kelam, 2015, p. 13). A similar pattern is visible in synthetic biology. Any new method or discovery that has economic value gets patent protection. Moreover, as with genetically modified crops, the whole point boils down to pursuing a for-profit and a monopoly on life.

References

Arber, W. (2014). Horizontal gene transfer among bacteria and its role in biological evolution. *Life*, *4*(2), 217–224. https://doi.org/10.3390/life4020217

Armstrong, R., Schmidt, M., & Bedau, M. (2012). Other developments in synthetic biology. In M. Schmidt (Ed.), *Synthetic biology: Industrial and environmental applications* (pp. 145–156). Wiley-Blackwell.

132 Ivica Kelam

Atlas, R. M., & Dando, M. (2006). The dual-use dilemma for the life sciences: Perspectives, conundrums, and global solutions. *Biosecurity and Bioterrorism: Biodefense Strategy, Practice, and. Science, 4*(3), 276–286. https://doi.org/10.1089/bsp.2006.4.276

Ball, P. (2004). Synthetic biology: Starting from scratch. *Nature, 431*(7009), 624–626. https://doi.org/10.1038/431624a

Beacham, T. A., Sweet, J. B., & Allen, M. J. (2017). Large scale cultivation of genetically modified microalgae: A new era for environmental risk assessment. *Algal Research, 25*, 90–100. https://doi.org/10.1016/j.algal.2017.04.028

biofuelwatch.org. (2017). Microalgae biofuels myths and risks. https://www.biofuelwatch.org.uk/wp-content/uploads/Microalgae-Biofuels-Myths-and-Risks-FINAL.pdf. Accessed on July 21, 2023.

Bugaj, L. J., & Schaffer, D. V. (2012). Bringing next-generation therapeutics to the clinic through synthetic biology. *Current Opinion in Chemical Biology, 16*(3–4), 355–361. https://doi.org/10.1016/j.cbpa.2012.04.009

Bügl, H., Danner, J. P., Molinari, R. J., Mulligan, J. T., Park, H.-O., Reichert, B., Roth, D. A., Wagner, R., Budowle, B., Scripp, R. M., Smith, J. A. L., Steele, S. J., Church, G., & Endy, D. (2007). DNA synthesis and biological security. *Nature Biotechnology, 25*(6), 627–629. https://doi.org/10.1038/nbt0607-627

Burnett, L., Lunn, G., & Coico, R. (2009). Biosafety: Guidelines for working with pathogenic and infectious microorganisms. *Current Protocols in Microbiology, 13*(1), 1A.1.1–1A.1.14. https://doi.org/10.1002/9780471729259.mc01a01s13

Callaway, E. (2016). "Minimal" cell raises stakes in race to harness synthetic life. *Nature, 531*(7596), 557–558. https://doi.org/10.1038/531557a

Cameron, E. D., Bashar, C. J., & Collins, J. J. (2014). A brief history of synthetic biology. *Nature Reviews Microbiology, 12*(5), 381–390. https://doi.org/10.1038/nrmicro3239

Carter, S. R. C., Rodemeyer, M., Garfinkel, M. S., & Friedman, R. M. (2014). Synthetic biology and the U.S. Biotechnology Regulatory System: challenges and options. https://www.wiley.law/media/publication/279_full-report.pdf. Accessed on July 21, 2023.

Church, G. M., Elowitz, M. B., Smolke, C. D., Voigt, C. A., & Weiss, R. (2014). Realizing the potential of synthetic biology. *Nature Reviews Molecular Cell Biology, 15*(4), 289–294. https://doi.org/10.1038/nrm3767

Convention on Biological Diversity. (2016). Conference of the parties to the convention on biological diversity. "Decision Adopted by the Conference of the Parties to the Convention on Biological Diversity XIII/17". *Synthetic Biology*. https://www.cbd.int/doc/decisions/cop-13/cop-13-dec-17-en.pdf. Accessed on July 21, 2023.

Danino, T., Prindle, A., Kwong, G. A., Skalak, M., Li, H., Allen, K., Hasty, J., Bhatia, S. N. (2015). Programmable probiotics for detection of cancer in urine. *Science Translational Medicine, 7*(289), 289ra84. https://doi.org/10.1126/scitranslmed.aaa3519

Davison, J. (1999). Genetic exchange between bacteria in the environment. *Plasmid, 42*(2), 73–91. https://doi.org/10.1006/plas.1999.1421

De Vriend, H. (2006). *Constructing Life. Early social reflections on the emerging field of synthetic biology*. Rathenau Institute. Working Document 97. https://www.rathenau.nl/sites/default/files/2018-05/boek_WED97_Constructing_Life_2006.pdf. Accessed on July 21, 2023.

Dormitzer, P. R., Suphaphiphat, P., Gibson, D. G., Wentworth, D. E., Stockwell, T. B., Algire, M. A., Alperovich, N., Barro, M., Brown, D. M., Craig, S., Dattilo, B. M., Denisova, E. A., De Souza, I., Eickmann, M., Dugan, V. G., Ferrari, A., Gomila, R. C., Han, L., Judge, C., ... Venter, J. C. (2013). Synthetic generation of influenza vaccine viruses for rapid response to pandemics. *Science Translational Medicine, 5*(185), 185ra68. https://doi.org/10.1126/scitranslmed.3006368

Douglas, T., & Savulescu, J. (2010). Synthetic biology and the ethics of knowledge. *Journal of Medical Ethics, 36*(11), 687–693. https://doi.org/10.1136/jme.2010.038232

Dröge, M., Pühler, A., & Selbitschka, W. (1998). Horizontal gene transfer as a biosafety issue: A natural phenomenon of public concern. *Journal of Biotechnology, 64*(1), 75–90. https://doi.org/10.1016/s0168-1656(98)00105-9

EnAlgae Consortium. (2015). EnAlgae in conclusion: Products and impacts. http://www.enalgae.eu/public-deliverables.htm. Accessed on July 21, 2023.

European Commission Scientific Committees. (2014). Opinion on synthetic biology I risk assessment methodologies and safety aspects. https://ec.europa.eu/health/scientific_committees/emerging/docs/scenihr_o_044.pdf. Accessed on July 21, 2023.

Ferber, D. (2004). Microbes made to order. *Science, 303*(5655), 158–161. https://doi.org/10.1126/science.303.5655.158

Flynn, K. J., Mitra, A., Greenwell, H. C., & Sui, J. (2012). Monster potential meets potential monster: Pros and cons of deploying genetically modified microalgae for biofuels production. *Interface Focus, 3*(1), 20120037. https://doi.org/10.1098/rsfs.2012.0037

FOE, ICTA, & The ETC Group. (2012). Principles for the oversight of synthetic biology. https://www.etcgroup.org/sites/www.etcgroup.org/files/The%20Principles%20for%20the%20Oversight%20of%20Synthetic%20Biology%20FINAL.pdf. Accessed on July 21, 2023.

Garfinkel, M. S., Endy, D., Epstein, G. L., & Friedman, R. M. (2007). *Synthetic biology: Options for governance.* Center for Strategic and International Studies. MIT.

Getino, M., Sanabria-Ríos, D. J., Fernández-López, R., Campos-Gómez, J., Sánchez-López, J. M., Fernández, A., & de la Cruz, F. (2015). Synthetic fatty acids prevent plasmid-mediated horizontal gene transfer. *mBio, 6*(5). https://doi.org/10.1128/mbio.01032-15

Gibson, D. G., Glass, J. I., Lartigue, C., Noskov, V. N., Chuang, R.-Y., Algire, M. A., & Venter, J. C. (2010). Creation of a bacterial cell controlled by a chemically synthesized genome. *Science, 329*(5987), 52–56. https://doi.org/10.1126/science.1190719

Goldenfeld, N., & Woese, C. (2007). Biology's next revolution. *Nature, 445*(7126), 369. https://doi.org/10.1038/445369a

Grunwald, A. (2012). Synthetic biology: Moral, epistemic and political dimensions of responsibility. In R. Paslack, J. S. A. Ach, B. Luttenberg, & K. M. Weltring (Eds), *Proceed with caution?: Concept and application of the precautionary principle in nanobiotechnology* (pp. 243–259). LIT Verlag.

Guan, Z., Schmidt, M., Pei, L., Wei, W., & Ma, K. (2013). Biosafety considerations of synthetic biology in the International Genetically Engineered Machine (iGEM) Competition. *BioScience, 63*(1), 25–34. https://doi.org/10.1525/bio.2013.63.1.7

134 Ivica Kelam

Hobom, B. (1980). Surgery of genes-at the doorstep of synthetic biology. *Medizinische Klinik, 75*(24), 14–21.

Howell, K. (2009, July 14). ExxonMobil bets $600 million on algae. *Scientific American.* http://www.scientificamerican.com/article.cfm?id=biofuels-algae-exxonventer. Accessed on July 21, 2023.

Hutchison, C. A., Chuang, R.-Y., Noskov, V. N., Assad-Garcia, N., Deerinck, T. J., Ellisman, M. H., & Venter, J. C. (2016). Design and synthesis of a minimal bacterial genome. *Science, 351*(6280), aad6253. https://doi.org/10.1126/science.aad6253

ieabioenergy.com. (2017). State of technology review – Algae Bioenergy - An IEA Bioenergy Inter-Task Strategic Project. https://www.ieabioenergy.com/wp-content/uploads/2017/02/IEA-Bioenergy-Algae-report-update-Final-template-20170131.pdf. Accessed on July 21, 2023.

Kelam, I. (2015). Genetički modificirani usjevi kao bioetički problem. *Genetically modified crops as a bioethical issue.* Zagreb/Osijek: Pergamena/Visoko evanđeosko teološko učilište/Centar za integrativnu bioetiku.

Kelle, A. (2012). Beyond Patchwork Precaution in the Dual-Use Governance of Synthetic Biology. *Science and Engineering Ethics, 19*(3), 1121–1139. https://doi.org/10.1007/s11948-012-9365-8

Kelle, A. (2013). Synthetic biology as a field of dual-use bioethical concern. In B. Rappert & M. J. Selgelid (Eds.), *On the Dual Uses of Science and Ethics: Principles, Practices, and Prospects Paperback* (Vol. 2013, pp. 45–64). ANU Press.

Leduc, S. (1910). Théorie Physico-chimique de la Vie et Générations Spontanées. Paris: A. Poinat. https://archive.org/details/thoriephysicoc00leduuoft/page/n9/mode/2up?view=theater. Accessed on July 21, 2023.

Leduc, S. (1912). *La Biologie Synthetique.* A. Poinat. http://www.peiresc.org/bstitre.htm. Accessed on July 21, 2023.

Lewis, R. (2013). How Craig Venter created life. https://dnascience.plos.org/2013/10/10/how-craig-venter-created-life/. Accessed on July 21, 2023.

Marris, C. (2015). The construction of imaginaries of the public as a threat to synthetic biology. *Science as Culture, 24*(1), 83–98. https://doi.org/10.1080/09505431.2014.986320

Maurer, S. M., & Zoloth, L. (2020). Synthesizing Biosecurity. *Bulletin of the Atomic Scientists, 63*(6), 16–18. https://doi.org/10.1080/00963402.2007.11461114

National Bioeconomy Blueprint. (2012). *Washington: The White House.* https://obamawhitehouse.archives.gov/sites/default/files/microsites/ostp/national_bioeconomy_blueprint_april_2012.pdf. Accessed on July 21, 2023.

National Research Council. (2012). *Sustainable Development of Algal Biofuels in the United States* (p. 2). The National Academies Press. https://doi.org/10.17226/13437. Accessed on July 21, 2023.

Official Journal of the European Union. (2018). "COMMISSION DIRECTIVE (EU) 2018/350 of 8 March 2018 amending Directive 2001/18/EC of the European Parliament and of the Council as regards the environmental risk assessment of genetically modified organisms." https://eur-lex.europa.eu/legal-content/EN/TXT/PDF/?uri=CELEX:32018L0350&from=EN. Accessed on July 21, 2023.

Paddon, C. J., & Keasling, J. D. (2014). Semi-synthetic artemisinin: A model for the use of synthetic biology in pharmaceutical development. *Nature Reviews Microbiology, 12*(5), 355–367. https://doi.org/10.1038/nrmicro3240

Ethical Aspects of Promises and Perils **135**

Pamlin, D., & Armstrong, S. (2015). *12 Risks that threaten human civilization: The case for a new risk category*. Global Challenges Foundation. http://globalchallenges.org/wp-content/uploads/12-Risks-with-infinite-impact.pdf. Accessed on July 21, 2023.

Pennisi, E. (2010). Genomics. Synthetic genome brings new life to bacterium. *Science, 328*(5981), 958–959. https://doi.org/10.1126%2Fscience.328.5981.958

Peplow, M. (2016, February 25). Synthetic biology's first malaria drug meets market resistance. *Nature, 530*(7591), 389–390. https://doi.org/10.1038/530390a

President's Commission on the Study of Bioethical Issues. (2010). New directions: The ethics of synthetic biology and emerging technologies. *President's Commission on the Study of Bioethical Issues*. https://bioethicsarchive.georgetown.edu/pcsbi/sites/default/files/PCSBI-Synthetic-Biology-Report-12.16.10_0.pdf. Accessed on July 21, 2023.

Purnick, P. E. M., & Weiss, R. (2009). The second wave of synthetic biology: From modules to systems. *Nature Reviews Molecular Cell Biology, 10*(6), 410–422. https://doi.org/10.1038/nrm2698

Rawls, R. L. (2000). 'Synthetic Biology' makes its debut. *Chemical & Engineering News, 78*(17), 49–53. https://doi.org/10.1021/cen-v078n017.p049

Rojahn, S. Y. (2013). Synthetic biology could speed flu vaccine production. *MIT Technology Review* https://www.technologyreview.com/2013/05/14/178443/synthetic-biology-could-speed-flu-vaccine-production/. Accessed on July 21, 2023.

Rossi, F., Rizzotti, L., Felis, G. E., & Torriani, S. (2014). Horizontal gene transfer among microorganisms in food: Current knowledge and future perspectives. *Food Microbiology, 42*, 232–243. https://doi.org/10.1016/j.fm.2014.04.004

Royal Academy of Engineering. (2009). *Synthetic biology: scope, applications, and implications*. Royal Academy of Engineering. https://www.cbd.int/doc/emerging-issues/UK-submission-2011-013-Synthetic_biology-en.pdf. Accessed on July 21, 2023.

Ruder, W., Lu, T., & Collins, J. (2011). Synthetic Biology Moving into the Clinic. *Science, 333*(6047), 1248–1252. https://doi.org/10.1126/science.1206843

Rugnetta, M. (2024). Synthetic biology. https://www.britannica.com/science/synthetic-biology. Accessed on July 21, 2023.

Sample, I. (2010, May 20). Craig Venter creates synthetic life form. *Guardian*. https://www.theguardian.com/science/2010/may/20/craig-venter-synthetic-life-form. Accessed on July 21, 2023.

Schmidt, C. W. (2010). Synthetic biology. Environmental health implications of a new field. *Environmental Health Perspectives, 118*(3), A118–A123. https://doi.org/10.2307/25615015.

Selgelid, M. J. (2007). A tale of two studies: ethics, bioterrorism, and the censorship of science. *Hastings Center Report, 37*(3), 35–43. https://doi.org/10.1353/hcr.2007.0046

Sharma, Y. (2012). NGOs call for international regulation of synthetic biology. https://www.scidev.net/global/genomics/news/ngos-call-for-international-regulation-of-synthetic-biology.html. Accessed on July 21, 2023.

Singh, M., & Vaidya, A. (2015). Translational synthetic biology. *Systems and Synthetic Biology, 9*(4), 191–195. https://doi.org/10.1007/s11693-015-9181-y

Szybalski, W. (1974). In Vivo and in Vitro initiation of transcription, page 405. In A. Kohn & A. Shatkay (Eds), *Control of Gene Expression*. Plenum Press. 23–4, and Discussion pp. 404–5 (Szybalski's concept of Synthetic Biology), 411–2, 415–7.

136   Ivica Kelam

Szybalski, W., & Skalka, A. (1978). Nobel prizes and restriction enzymes. *Gene*, *4*(3), 181–182. https://doi.org/10.1016/0378-1119(78)90016-1

Tang, W. H. W., Wang, Z., Levison, B. S., Koeth, R. A., Britt, E. B., Fu, X., Wu, Y., & Hazen, S. L. (2013). Intestinal microbial metabolism of phosphatidylcholine and cardiovascular risk. *New England Journal of Medicine*, *368*(17), 1575–1584. https://doi.org/10.1056/nejmoa1109400

Thomas, C. M., & Nielsen, K. M. (2005). Mechanisms of and barriers to, horizontal gene transfer between bacteria. *Nature Reviews Microbiology*, *3*(9), 711–721. https://doi.org/10.1038/nrmicro1234

Trump, B. D. (2020). Synthetic biology, GMO, and risk: What is new, and what is different? In B. D. Trump, C. L. Cummings, J. Kuzma, & I. Linkov (eds), *Synthetic Biology 2020: Frontiers in Risk Analysis and Governance* (pp. 85–105). Springer.

Tucker, J. B., Zilinskas, R. A. (2006). The promise and perils of synthetic biology. *The New Atlantis*, *12*: 25–45. https://www.thenewatlantis.com/wp-content/uploads/legacy-pdfs/TNA12-TuckerZilinskas.pdf. Accessed on July 21, 2023.

Van Est, R., De Vriend, H., & Walhout, B. (2007). *Constructing life the world of synthetic biology*. Rathenau Institute. https://haseloff.plantsci.cam.ac.uk/resources/SynBio_reports/BAP_Synthetic_biology_nov2007%5B1%5D.pdf. Accessed on July 21, 2023.

Voosen, P. (2013). Synthetic biology comes down to earth. *The Chronicle of Higher Education*. https://www.chronicle.com/article/synthetic-biology-comes-down-to-earth/. Accessed on July 21, 2023.

Washburn, J. (2007, March 24). Big oil buys Berkeley. *Los Angeles Times*. https://www.latimes.com/la-oe-washburn24mar24-story.html. Accessed on July 21, 2023.

White, K., & Vemulpad, S. (2015). Synthetic biology and the responsible conduct of research, *Macquarie Law Journal*, *15*, 59–63. https://search.informit.org/doi/10.3316/informit

Wright, O., Stan, G. B., & Ellis, T. (2013). Building-in biosafety for synthetic biology. *Microbiology*, *159*(pt 7), 1221–1235. https://doi.org/10.1099/mic.0.066308-0

Chapter 7

Adapting (Bio)ethics to Technology and Vice Versa: When to Fight and When to Collaborate With Artificial Intelligence

Iva Rinčić and Amir Muzur

University of Rijeka, Croatia

Abstract

The rapid advancement of artificial intelligence (AI), particularly within the last decade and the application of 'deep learning', has simultaneously accelerated human fears of the changes AI provokes in human behaviour. The question is not any more if the new phenomena, like artificially-induced consciousness, empathy or creation, will be widely used, but whether they will be used in ethically acceptable ways and for ethically acceptable purposes.

Departing from a diagnosis of the state humans have brought themselves to by (ab)use of technology, the present chapter investigates the possibility of a systematic study of adaptations human society will have to consider in order to guarantee the obeyance to the fundamental ethical values and thus its spiritual survival. To that end, a new discipline – epharmology (from the Greek epharmozein = to adapt) is proposed, together with its aims and methodology.

Keywords: Bioethics; AI; epharmology; deep learning'; empathy

When Did Things Go Wrong? A Brief History of AI

Already in the first half of the 20th century, the Czech writer Karel Čapek (1890–1938), nominated seven times for the Nobel Prize, with the play R.U.R. (Rossum's Universal Robots) from 1920 spread the term 'robot' around the world (the term was suggested to him by his brother, Josef Čapek, 1887–1945, painter and writer who died in a concentration camp). Although it seems prophetic today,

---

The Ethics Gap in the Engineering of the Future, 137–156

Copyright © 2025 Iva Rinčić and Amir Muzur

Published under exclusive licence by Emerald Publishing Limited

doi:10.1108/978-1-83797-635-520241008

138 Iva Rinčić and Amir Muzur

Čapek's notion must have looked like a flimsy fiction unworthy of fiction exactly a 100 years ago. After all, what did the world of that time even know about the ideologies, dictatorships and the reverse of progress that it would soon meet and that would change it so thoroughly?

Today, many see the beginnings of the history of AI only after the Second World War – primarily in the work of Alan Turing (1912–1954), the 'Enigma' decipherer. (That AI could not be significantly developed before 1949 is also shown by the fact that until then computers did not have the ability to store commands at all but only to execute them.)[1] Back in the 1930s, Turing put forward a thesis about the so-called Turing's universal machine, and in 1950, he designed the so-called Turing test of 'machine intelligence', depending on whether a person can distinguish the answers of a computer or a person. Vannevar Bush, the wartime head of the US Office of Scientific Research and Development, in a 1945 article entitled 'As we may think', announces a machine called Memex as an 'addition to one's own memory'.

In 1950, Isaac Asimov (1920–1992) published a collection of stories 'I, Robot' (I, Robot), in which he advocated the 'Three Laws of Robotics': a robot must not hurt a human or allow passivity to man is hurt; he must obey the orders of men unless such orders would conflict with the first law; he must protect his own existence as long as it does not conflict with the first two laws. No matter how unrealistic they may seem at the time, these 'laws' are a pioneering attempt at robot ethics.

At a conference held in 1956 at Dartmouth University, mathematician John McCarthy gathered three other participants – Marvin Minsky from Harvard and two experts from Bell Telephone Laboratories (IBM) – coining the term 'AI', they predicted that 'the work will be based on the assumption that any aspect of learning or any other characteristic of intelligence can in principle be described so precisely that it can be simulated by a machine'.

Hopes about the imminent invention of intelligent machines quickly proved to be too optimistic: the first crisis (from 1974 to the early 1980s) and then the second (1987–1993) of AI funding occurred. Interest (and fear) certainly revived with the news that the current chess champion Garry Kasparov was defeated by the IBM machine 'Deep Blue' in 1997.[2]

In the same year, Dragon Systems from Massachusetts launched the first successful voice recognition programme (Naturally Speaking), and in the 1990s, Cynthia Beazeal from MIT introduced the emotion-recognizing robot Kismet.[3]

In 2005, the American inventor Raymond Kurzweil predicted that around 2045 AI will threaten human intelligence (the so-called singularity); however,

---

[1]Anyoha, R. (2017, July 2). The history of Artificial Intelligence. *Science in the News* (SITN, Special Edition Summer). https://sitn.hms.harvard.edu/flash/2017/history-artificial-intelligence/.

[2]Tate, K. (2014, June 3). *History of A. I.: Artificial Intelligence.* https://www.livescience.com/47544-history-of-a-i-artificial-intelligence-infographic.html.

[3]As expected, more perfect versions of 'social robots', such as Pepper (2014) or Sofia (2016), will receive even greater attention from the media around the world.

Adapting (Bio)ethics to Technology and Vice Versa   *139*

Marvin Minsky announced the same thing in 1970 in '3-8 years'[...] In support of Kurzweil's assessment, in 2011 the Apple company introduced a 'personal assistant' – Siri – on its iPhone 45, and even more so when in 2012 'deep learning' (deep learning, actually learning based on experience) took off suddenly, mainly in the service of visual recognition and speech analysis, which at the end of 2022 introduced the famous ChatGPT into wider use. The repercussions of this invention (that is, in a broader sense, deep learning) are so great that any previous history of AI has lost its significance. A 2006 history of AI[4] limits the main research advances of the second half of the 20th century to search algorithms, machine learning and statistical analysis. In short, AI is credited with helping solve spam blocking, image and voice recognition that it greatly improved the search, but that, in relation to Turing's forecasts, it completely failed. When it comes to AI, in fact, everything we fear, everything we hope for, is based on a 10-year-old milestone.[5,6]

What Are We Really Afraid of?

Judging by the current discourse, are we fighting for advantage, supremacy, justice or survival?

The American writer Howard Phillips Lovecraft (1890–1937) could not have guessed that in 2023 his 'Shoggoth' would be accepted as a symbol of human fears before a new generation of artificially intelligent computer programs. The panic reached far greater proportions in March 2023, when 27,000 AI researchers and technologists signed an appeal demanding a multi-month moratorium on the development, production and use of the so-called of large language models (LLMs), which represent a 'deep risk for society and humanity'. Many claimed that their signatures on the support appeal were falsified, and there was a sus-picion that it was the selfish interest of some producers who would gain an advantage through the appeal. In any case, the key question is: what real danger does a programme such as ChatGPT (Chat Generative Pre-Trained Transformer, i.e. 'chatbot', i.e. 'chattering robot' released on the market in November 2022 by OpenAI, which simulates a conversation) represent and in 'dialogue' he acquires

---

[4]Smith, C., McGuire, B., Huang, T. & Yang, G. (2006, December). *The history of Artificial Intelligence.* https://courses.cs.washington.edu/courses/csep590/06au/projects/history-ai.pdf.
[5]A certain neurolinguist from Trieste claimed back in 1994 that a computer would never be invented that would be able to distinguish the nuances of the individual melody of the human voice.
[6]'Deep learning', as a subtype of machine learning, has been known to some extent probably since the 1940s and the first attempts to construct artificial neurons and neural networks. The Finnish mathematician Seppo Linnainmaa in the 1960s, and then the Canadian Hinton and the American Rumelhart, invented the retrograde propagation algorithm, and by the end of the 1990s further progress was made in imitating image recognition and long-term memory. However, the turning point in 2012, carried out by Hinton and colleagues, at the same time as the Google research team, led to a real shift whose reach we still cannot estimate.

knowledge using the deep learning method)? Objections to it are mainly based on three 'predictions': the long-term atrophy of human intelligence (since ChatGPT talks, writes and takes the place of humans), the spread of false information (texts, photos, images, videos) and, finally, the displacement of a whole range of professions that will be replaced by chatbots (translators, lawyers, radiologists, etc.). About these 'dangers' are pointed out by Geoffrey Hinton himself, one of the pioneers of deep learning, who left Google in April 2023. Of course, it can be said that ChatGPT is not ready yet, that its results turn out to be false (to the point of comedy) due to incompleteness and immaturity programme (this is pointed out, for example, by the Pennsylvanian bioethicist Jon Merz, while Arthur Caplan and Lee Igel think that it would be irresponsible not to use AI but also to help free practitioners from administrative tasks or to improve the availability of knowledge to a wider circle).

However, in order to understand the problems of science, education and human creation, we need to reach deeper than the current crisis related to digital technology.

The First Problematic Intermezzo: Consciousness and AI[7]

The lack of a clear definition of consciousness inevitably makes it difficult to contextualize it in relation to AI. Consciousness has been defined differently: as 'choice' (selecting agency; William James), 'state of insight' (Philip Zimbardo), 'integration and stabilization' (Gordon Allport), 'mutual congruence' (Marcel Kinsbourne), 'control in relation to automatism' (Carlo Umiltà), 'the behaviour of a huge number of neurons and the molecules connected to them' (Francis Crick), etc. Although this review of definitions is far from complete, it is almost certain that AI meets the criterion of stabilization and integration (in which even better than man), efficient neural networks (that is, an artificial equivalent of neurons), differentiating 'controlled' from 'automatic' (at two levels of reaction), probably has a certain choice (perhaps something like human 'programed freedom'), but, in its opportunity, does not have the real 'insight' that Zimbardo insists on. The Shallice team, as early as 40 years ago, emphasized behaviour monitoring, movement control, enabling flexibility in response and decision-making as the main features of consciousness: AI can be said to possess them all. Of the three levels of consciousness that Tulving highlighted in the mid-1980s - anoetic (about external stimuli), noetic (about symbolic representations of the world) and autonoetic (about oneself and one's own experience) - AI, it can be assumed, certainly has the first, maybe the second, but most likely not the third. Recently, there is talk of 'strong AI' ('artificial general intelligence'), which implies consciousness (subjective experience and thought), self-awareness (awareness of oneself as a separate individual), sensitivity and wisdom. There is no doubt, however, that consciousness means responsibility, from which the question follows:

---

[7]Some ideas included in this subtitle were previously presented at the Second International Conference on Artificial Intelligence Humanities ICAIH 2019, Seoul, R. Korea, 14 August 2019, under the title *Artificial intelligence and consciousness: paving the way for an epharmological analysis*.

Adapting (Bio)ethics to Technology and Vice Versa 141

Can AI be held accountable for its actions? Consciousness implies certain rights: should they also be provided to artificially intelligent machines (as, more or less successfully, we provide them to animals, and we also think about plants[8])?

In any consideration of consciousness related to AI, of course, it is possible to (mis)use John Searle's famous thought experiment from 1980 - the so-called 'Chinese speaking room.' And so we come to the key question: can consciousness (at all) be programed? Because if it can't, consciousness in AI is an illusion. But if it can, then AI awareness is only a matter of time. In plants, learning and memory (habituation), active movement (flowering, leaf closure), exploration of the environment directed by networked roots, etc. are described. Animals and humans are born with instincts, while other functions (including the most complex ones, such as morality) are learnt (while speech could be one of the unprogrammed, evolutionarily conditioned functions). Can learning and adaptation in AI cause the emergence of unprogrammed functions? In any case, learning and adaptation can be considered 'programmable' elements of consciousness (e.g. the 'Deep Blue' chess computer: where, of course, in humans it is possible to increase memory with contextual information, and in AI – with data compression), as well as empathy. It is theoretically possible to programme both sensation (pain: will it determine our attitude about AI rights, as it determined the attitude about animal rights?) and morality (as a variation of learning and adaptation (to socially acceptable and unacceptable behaviour). The hardest thing to imagine programming of freedom, initiative, insight into oneself, self-sufficiency (thus, anti-zombie behaviour that is characterized by independence from external conditions and is capable of resulting in unexpected products) or abstract concepts (like art or science) that are not an immediate feedback loop in relation to that which it produces, but are important for others (thus, for wider aesthetics) - here we are not talking about the predominant products of today's 'science', which has turned into satisfying the prescribed form and imitating patterns, but about the original ingenious science aimed at the common good. in other cases, and in the consideration of consciousness, the emergence of AI has taught us to ask different, perhaps more important questions.

Another Problematic Intermezzo: Empathy and AI[9]

One of the sub-questions of the possibility of developing consciousness in AI is the development of empathy as an important feature of consciousness.

David Hume (1739) equated 'sympathy' with 'affective empathy', which Robert Vischer (1873) would call 'enjoyment' (Einfühlung) and connect it with

---

[8]In October 2017, Saudi Arabia granted citizenship to a robot named Sofia who 'can see, understand conversation and create relationships', and a month later the talking robot ('seven-year-old boy') – chatbot Mirai was legally granted residency in Tokyo.

[9]Some of the ideas included in this subtitle were presented at the Third International Conference on Artificial Intelligence Humanities ICAIH 2020, Seoul, R. Korea (online), December 8–14, 2020 under the title *Artificial Intelligence and Empathy: Living in Perfect Harmony?*

142   Iva Rinčić and Amir Muzur

aesthetics, and Theodor Lipps (1897) with understanding another's mind (1909). Edward Titchener will translate the German Einfühlung into English as 'empathy'. Adam Smith asserted in 1759 that empathy, unlike pity, has a communicative role. According to Alfred Kaszniak, compassion is empathy associated with another's suffering, increased by the desire to reduce that suffering, while Matthew Ricard sees compassion as empathy combined with altruistic love, emphasizing that empathy can experience fatigue, but compassion cannot. According to more modern research (Martin Hoffman Karsten Stueber etc.), empathy is an affective response that reflects another's situation and imagines another's thought processes, or, as Cristina Gonzales-Liencres et al. say, empathy allows an individual to share affective states with others, predict others' actions and encourage prosocial behaviour.[10–12]

In animals where it has been observed, such as primates, elephants or dolphins, empathy is associated with activity in the anterior cingulum and insula.[13] When observing other people's suffering, 'mirror' neurons are also activated. The ability to communicate emotions (their recognition and expression), regulation of emotions (modelled by social, psychological and biological factors such as attention, mood, etc.), and cognitive mechanisms of emotion assessment are mentioned as components of the empathy model.[14] Empathy has similarities with emotions (combination of mental and cognitive activity, simultaneous activation of facial expressions, bodily reactions and vocalizations, etc.), but it is not clear whether empathy, by its nature, is an emotion or a reflex or both.[15]

The connection of empathy with morality is not completely clear either, although it is more often seen as a key force of morality.[16]

---

[10]Hoffman, M. L. (2000). *Empathy and moral development: Implications for caring and justice.* Cambridge University Press.

[11]Stueber, K. (2006). *Rediscovering empathy: Agency, folk psychology, and the human sciences.* MIT Press.

[12]Gonzalez-Liencres, C., Shamay-Tsoory, S. G., & Brüne, M. (2013). Towards a neuroscience of empathy: ontogeny, phylogeny, brain mechanisms, context and psychopathology. *Neuroscience and Biobehavioral Reviews, 37*(8), 1537–1548.

[13]For deeper insight on the evolutionary perspective of empathy, see: Decety, J. (2011). The neuroevolution of empathy. *Annals of the New York Academy of Sciences, 1231(1),* 35–45; Decety, J., Norman, G. J., Berntson, G., & Cacioppo, J. T. (2012). A neurobehavioral evolutionary perspective on the mechanisms underlying empathy. *Progress in Neurobiology, 98*(1), 38–48.

[14]On the differences between affective and cognitive component disorders, see Bartochowsky, Z., Gatla, S., Khoury, R., Al-Dahhak, R., & Grossberg, G. T. (2018). Empathy changes in neurocognitive disorders: A review. *Annals of Clinical Psychiatry, 30*(3), 200–232.

[15]Further readings in Yalçin, O. N., & DiPaola, S. (2020). Modeling empathy: building a link bween affective and cognitive processes. *Artificial Intelligence Review, 53*(4), 2983–3006.

[16]Heyes, C. (2018). Empathy is not in our genes. *Neuroscience and Biobehavioral Reviews, 95,* 499–507.

Adapting (Bio)ethics to Technology and Vice Versa   *143*

The cultural colouring of empathy is more certain: Americans, for example, compared to the Japanese, report more intense and longer emotions, and the British perceive needle pricks and other people's suffering more negatively than East Asians.[17] And time plays a role in the development of empathy: the average American student in 2009 turned out to be less empathetic than most students tested in 1979.[18] Numerous studies have registered a decline in empathy or moral reasoning during medical school.[19,20]

Given the general challenges of programming consciousness in AI, it's no wonder that 'designing artificial empathy [has] become one of the most pressing human-robot relationship issues', as Minoru Asada, professor of neurorobotics at Osaka University, says.[21] It is certain that even now robots can be programed to recognize facial expressions, body reactions and prosody (even in multiple languages). What will remain a more difficult task will be (learning) to recognize individual modulations of empathy under the influence of earlier experiences, current changes in mood and environment. (Considering that newborns can show signs of empathy, responding by crying to the crying of another child, it is not clear how much empathy is reflexive: it is possible that only its affective component is reflexive, and the cognitive component is acquired through learning or experience.) Help is announced with great fanfare. provide artificial empathy algorithms to social workers, predicting with over 90% probability the risk of suicide.[22]

Robots will be able to learn 'morality' as a system of socially acceptable rules, but it remains questionable how they will learn one of the most intriguing human skills - lying. In addition, there remains the question of intention (which plays a large role in considerations of religion, ethics and law): 'positive' (learnt) emotions could confuse or even hurt users, (robotic) care (although it is also questionable whether human behaviour is always truly altruistically motivated), and children could develop such

---

[17]Atkins, D., Uskul, A. K., & Cooper, N. R. (2016). Culture shapes empathic responses to physical and social pain. *Emotion, 16*(5), 587–601.

[18]Seppälä, E. (2019, June 11). Empathy is on the decline in this country: A new book describes what we can due to bring it back. *The Washington Post.* (https://www.washingtonpost.com/lifestyle/2019/06/11/empathy-is-decline-this-country-new-book-describes-what-we-can-do-bring-it-back/).

[19]Neumann, M., Edelhäuser, F., Tauschel, D., Fischer, M. R., Wirtz, M., Woopen, C., Haramati, A., & Scheffer, C. (2011). Empathy decline and its reasons: A systematic review of studies with medical students and residents- *Academic Medicine, 86*(8), 996–1009.

[20]Hren, D., Marušić, M., & Marušić, A. (2011). Regression of moral reasoning during medical education: Combined design study to evaluate the effect of clinical study years. *PLoS One, 6*(3), e17406.

[21]Asada, M. (2023, July 21). Affective developmental robotics: How can we design the development of artificial empathy? http://www.macs.hw.ac.uk/~kl360/HRI2014W/submission/S7.pdf.

[22]Artificial empathy (2023, June, 17). http://en.wikipedia.org/wiki/Artificial_empathy.

144   Iva Rinčić and Amir Muzur

false emotions through imitation.[23,24] (Not to mention that there are also those who study empathy towards robots, concluding that women have greater potential).[25] Among the many speculations that come to mind when thinking about artificial empathy, there are certainly encouraging expectations (a decrease in the number of rapes, paedophilia and sexual offences in general; a decrease in the loneliness of the elderly and sick; a decrease in the frequency of child neglect), but also those that cause anxiety (alienation between people, the decline in the frequency of marriages and births): as, after all, in all human inventions.

Higher Education and AI

Having respect to certain expectations, the vast majority of those who write about the modern education system, in any country, would agree that it is fundamentally ripe for thorough reform and that the problem is not only in teaching methods, but in his wrong paradigm burdened by general rigidity, inflexibility or aimlessness.[26–28] Here, however, we will first of all turn to the problem of methodology.

---

[23]Allen, C. (2018, February 7). How digitalisation of everything is making us more lonely. *The Conversation.* https://theconversation.com/how-the-digitalisation-of-everything-is-making-us-more-lonely-90870).

[24]In support of the idea of the importance of behavior, rather than the underlying drive, see: Baumgaertner, B., & Weiss, A. (2014). Do emotions matter in the ethics of human-robot interaction? Artificial empathy and companion robots. In *Proceedings of the third International Symposium on New Frontiers in Human Robot Interaction.* London: The Society for the Study of Artificial Intelligence and the Simulation of Behaviour. http://doc.gold.ac.uk/aisb50/AISB50-S19/AISB50-S19-Baumgaertner-paper.pdf. There are also studies that emphasize that empathy reflects individual motives in a certain context, and not the context itself: Weisz, E., & Zaki, J. (2018). Motivated empathy: A social neuroscience perspective. *Current Opinion in Psychology, 24,* 67–71.

[25]Chin, J. H., Haring K. S., & Kim, P. (2023). Understanding the neural mechanisms of empathy toward robots to shape future applications. *Frontiers in Neurorobotics, 17,* https://www.frontiersin.org/articles/10.3389/fnbot.2023.1145989/full.

[26]See discussion in Sesardić, N. (2022). *Konsenzus bez pokrića [Consensus without coverage].* Školska knjiga

[27]On the connection between the (multimedia didactic) method and didactic learning arrangements, see: Matijević, M., & Topolovčan, T. (2017). Izazovi i trendovi u multimedijskoj didaktici [Challenges and trends in multimedia didactics]. *Radovi Zavoda za znanstvenoistraživački i umjetnički rad u Bjelovaru, 11,* 87–99.

[28]Among the numerous criticisms, see a typical, though not necessarily the most accurate, one in: Zovko, V. (2016). ICT-enabled education – need for paradigm shift. *Croatian Journal of Education, 18*(2), 145–155. See also Prensky, M. (2023, May 28). *Digital natives, digital immigrants.* https://www.marcprensky.com/writing/Prensky%20-%20Digital%20Natives,%20Digital%20Immigrants%20-%20Part1.pdf.

Adapting (Bio)ethics to Technology and Vice Versa **145**

Although computers began to be used in education already in the 1950s (B. F. Skinner, Gordon Pask), more serious discussions and fears were only brought about by the widespread availability of ChatGPT in early 2023.[29] Namely, decades of experience with uncritical use and abuse of the Internet, overestimation of smartphones and tablets they compensated to some extent for the still current lack of theoretical and experimental argumentation that would define a measure that meets the purpose and at the same time prevent the dangers of excessive reliance on digital technology.[30] Forced 'distance learning' and confinement during the escalation of the COVID-19 pandemic, led most institutions to, more or less openly, reveal and acknowledge the unwanted effects of the 'new normal'.

ChatGPT, however, with its opening of the possibility of artificial generation of a written work (seminar paper, essay, knowledge test, final, diploma or doctoral thesis, 'scientific' article), encouraged a broader questioning of methods, goals, verification and the meaning of education in general. Here we will skip the usual (for companies that produce artificially intelligent machines or their advocates) advertising the application of AI in teaching as a solution for 'personalization of education', 'administrative relief of teachers', translation from foreign languages, statistics (big data) and forecasts based on it (passability, for example), etc., and we will concentrate on an important question: what does the unfettered application of AI in education mean for the future of human culture and intelligence?[31]

As it stands, we have to come to terms with the fact that AI, ostensibly offering assistance, has taken control of at least three common learning outcomes: memory (where the Internet has revealed itself as an 'external unit of memory'), empathy (as evidenced by early experiments such as the one with Pepper) and creativity (ChatGPT). As with all inventions that, at least in part, benefit humanity, bans are not an option: a review of learning outcomes should be considered. Could one return to teaching (and evaluating, as an outcome of teaching) intuition? Is intuition a gift that would lead us to harm others unjustifiably?

---

[29]For a more complete overview of the history of the application of AI in education, see: (2023). AI and education: rivals or allies? *Jahr – European Journal of Bioethics* (in print).

[30]See one of the reviews in: Nađ (2020). Sve se teže odvajamo od ekrana, ne vodeći računa o zdravlju: razgovor s Goranom Ivkićem [It's getting harder and harder to separate ourselves from the screen, not taking care of our health: a conversation with Goran Ivkić]. *Universitas*, 20–22.

[31]At the same time, we should not forget that, although before the COVID-19 pandemic, at three universities in Croatia, only 70% of teachers assessed themselves as digitally competent. Müller, M., & Varga, A. (2019). Digital competencies of teachers and associates at higher educational institutions in the Republic of Croatia. *Informatologia*, 52(1–2), 28–44. On the other hand, it is also interesting to find that there is no connection between the length of computer science classes and students' knowledge of this field: Mateš, L., Mladenović, M., & Mladenović, S. (2016). Znaju li studenti prve godine što je internet? [Do first-year students know what the Internet is?]. *Školski vjesnik: časopis za pedagogijsku teoriju i praksu*, 65, 105–117. About the cooperation of teachers and students during adaptations to digitization, see: Dimitriadi, Y. (2019). Who you're gonna call? The development of university digital leaders: A case study. *Media Studies, 10*(19), 102–118.

Perhaps the new era of AI will also lead us to question how the brain works (in light of insights from deep learning), what chess is (memory + combining + anticipation as a variant of 'memorizing in advance?'), and, finally, what is knowledge (the inscription on the New York City Library has long emphasized 'useful' knowledge – as if there was something else), what are the urges to reach true knowledge (spiritual enlightenment) in relation to imposed, even harmful or useless? We will definitely have to redefine both the subject and teaching methods: in what can and what can't the machine replace us? If it has consciousness (as a by-product of activity, say), perhaps the boundary doesn't even exist. And the term 'originality' needs to be redefined: anyway, probably everything we write (and, more generally, create) is taken from some of our old readings, we just forgot about them, or tried to forget them, to look more original. AI, however, with impeccable 'memory' (there is no forgetting for AI), but also unscrupulous regarding the question of authorship and authority, will subjugate our mind with 'knowledge' – a creation devoid of ethical elements and repercussions, far enough away to be unrecognizable, and at the same time in every respect someone else's. Will we learn to evaluate ideas? Will ethics (p)remain the only boundary between us and AI (at least in education)? And what if AI 'learns' to be ethical?

Epharmology: A New Science Based on Precaution[32]

Although there are already some suggestions on how to coexist with technology while preserving human nature (that is, its key cognitive capacities, social inter-action and feelings and ethics), for example, by applying decision theory, strengthening self-confidence, teaching academic integrity and responding to problems of dependence on technology by practicing impulse control, the truth is that, in fact, we are still not clear which of the tools and processes we use are useful and which are harmful and to what extent.[33]

The processes to which we are exposed (which ends with AI) can be reduced to digitalisation/digitisation (increased storage of binary coded information; increased use of computer technology and electronic images; merging of the real and virtual worlds; automated and networked system of virtual operations incomprehensible to the human mind by its spread, speed and number of com-binations), informatization (increased reliance on computers; increased produc-tion of and exposure to information), automation/robotization/AI (increased transformation of human-controlled processes into 'self-controlled' processes) and globalization (increased planetary interconnection of people and things).[34]

---

[32]The ideas are partly presented in Muzur, A., Rinčić, I, Shim, J., & Byun, S. (2020). Epharmology: A plea for a new science and a new education paradigm. *Nova Prisutnost, 18*(1), 39–46.

[33]Yamamoto, Y., & Ananou, S. (2015). Humanity in the digital age: Cognitive, social, emotional, and ethical implications. *Contemporary Educational Technology, 6*(1), 1–18.

[34]Kagermann, H. (2015). Change through digitization – value creation in the age of Industry 4.0. In: H. Albach, H. Meffert, A. Pinkwart & Ralf Reichwald (eds.), *Management of Permanent Change* (pp. 23–45). Springer Fachmedien.

Adapting (Bio)ethics to Technology and Vice Versa   147

Among the general adaptations to these processes, forced specialization (equally visible in science, education and practice) and quantitative selection instead of critical (qualitative) selection stands out. Information becomes 'knowledge' - right or wrong, deep or superficial, essential or irrelevant ('democratic knowledge'). Changes in human mental functioning and behaviour thus arise due to formal characteristics (enormity, diversity, speed), and not substantive ones: digitalisation accelerates processes (transfer of information, execution of orders, etc.) and increases the exploitation (spending) of resources (contrary to the concept of 'sustainable development'). The fact is that the average adult spends three hours a day online (some teenagers and 9 hours with different digital media: the recent COVID-19 pandemic has only worsened the situation).[35,36]

The list of consequences is hard to overstate: multi-tasking can cause information to bypass the hippocampus and go to the striatum (resulting in procedural instead of explicit memory); 'cognitive offloading' occurs (using technology even when it is not necessary); reading a PDF on a digital platform gives different results than reading a printed text (focus on more details vs. on an abstract level); we are attracted to the new instead of the important, etc. In short, the total availability of information outside of our own memory, namely the same (uncritical) availability (and aggressive offer) of valuable (evidence-based) and worthless (wrong or irrelevant) information, as well as the pace of information availability (e.g. PPT slides, where the consumer does not follow their gradual creation, as when reading, writing on the board, etc.), must have led to neurophysiological and psychological changes. The important question is which adaptations are important for society, the human race, the environment, and which are irrelevant? Which adjustments are short-term and which are long-term? Which adaptations are correct, positive, useful, and which are wrong, negative, harmful? Which adaptations must we avoid and which ones should we encourage? What adaptations can preserve significant human strengths (dreaming/ daydreaming, creativity/innovation, intuition, etc.), relying on the passive search for statistical correlation, typical of AI?[37]

Which discipline can answer these questions? Theoretical, experimental and clinical neuroscience (neuroanatomy, neurophysiology, psychology, neurology, psychiatry, etc.) could tell what is 'normal' and what is adaptive. Technical sciences, robotics, informatics, etc. which causes adaptations. Sociology/ anthropology – what it means for the individual and society. Ecology – what it means for the environment. Philosophy – what we want. Bioethics – what should be suggested. Law – which should be strictly regulated. History – what we have already encountered.

---

[35]Previously, it was recommended to limit children's access to the interface to two hours a day, and today it is suggested that parents supervise the use of digital media.

[36]Richtell, M. (2021, January 17). Children's screen time has soared in the pandemic, alarming parents and researchers. *The New York Times* https://www.nytimes.com/2021/01/16/health/covid-kids-tech-use.html.

[37]PricewaterhouseCoopers LLP (Pwc) CEO survey (2017, December). *Human value in the digital age.* https://www.pwc.nl/en/publicaties/human-value-in-the-digital-age.html.

Perhaps we need a new discipline, which would cover the existing ones: let's call it epharmology (Greek epharmozein = to adapt). This discipline would have the task of studying changes in the human brain, mind and behavior caused by digitization and computerization (internet, typing, mobile phones, SMS, etc.); evaluate the new dimensions of human capacities resulting from these changes (attention span, number of 'actions per minute', hyperactivity, etc.); predict further changes in the brain and, if they are negative, propose their prevention, and, if they are positive, propose how to better adapt to them; to assess the optimal biological profile of a person, proposing standards for improving the quality of life (how much sleep or screen time per day; whether movies should be subtitled or synchronized; how best to 'use' books or music, etc.); propose a program of radical reform of the educational system (content and methodology), followed by changes in the brain, mind and behavior in a way that will be maximally flexible and anticipatory. Methodologically, epharmology would collect information from all relevant sciences (generating an archive, library, online data repositories, etc.); processed information (systematic reviews, meta-analyses, etc.); encouraged interdisciplinary critical judgment of information; submitted proposals for new directions of research and proposals for ethical and legal positions.

When it comes to principles, epharmology would be guided by the principle that the maximum is not always the optimum; that there must always be a long-term perspective in mind; that a complete organism is formed by humans in combination with the environment and that 'regulation is always necessary when there is a possible risk to health, safety or the environment, even when the evidence supporting this is only speculative' (precautionary principle).

If we accept that life is the ability to grow and reproduce, learning is the acquisition of new facts and skills from the environment, and adaptation is learning that results in the integration of acquired content into mental and physical activity, epharmology would mean the promotion of only those adaptations that result in benefits for the individual without causing damage to the environment. In this light, the ideas of Joyce Schenkein and Sage Briggs (2016) should be viewed that 'adaptation is value neutral - we can become smarter or dumber, we just adapt to the environment. This means that we actually have more control over the influence of digital media than we think'.[38] that is, Antony Scriffigan (2017) that 'the challenge is to accept digital evolution, but to think about how it could potentiate marginalization, crowd out creativity, or in some other way lead to unwanted consequences.'[39]

Of course, the question is also what epharmology will look like in 10 years: will it mediate millions of court claims for damages from video game manufacturers (as happened to the tobacco industry in the 1990s), will it become an

---

[38]Briggs, S. (2016, September, 12). 6 ways digital media impacts the brain. *InformED*. https://www.opencolleges.edu.au/informed/features/5-ways-digital-media-impacts-brain/.
[39]https://www.linkedin.com/pulse/do-robots-viruses-get-sick-days-reflecting-impact-anthony.

Adapting (Bio)ethics to Technology and Vice Versa   *149*

indispensable body that consults governments, parliaments and industry? Or will it be declared mission impossible, since we've gone too far?[40]

The Ethics of Using AI

The ethics of AI has become a major challenge for many in recent years[41]

Some see privacy and surveillance, bias and discrimination, the question of trumping or not human judgement as the main problems of AI ethics.[42] 'Wise' companies create their own AI ethics, combining the principalism of mainstream bioethics ('respect for the person' = autonomy, beneficence, justice) with 'other issues' such as privacy (e.g. when processing large databases), pushing people out of work, bias and discrimination, etc. desiring that digital ethics directly 'mirror' medical ethics by formulating codes, ethical standards and respecting autonomy.[43,44,45] For others, 'explainability' (ie, the ability to trace data sources, results, algorithms, etc.) is primarily important.[46]) and responsibility (for the consequences of AI For third parties, it is only up to a declarative warning that AI is used (e.g. ChatGPT) or imitates a living artist, and the fourth ask AI to be what even humans are not – that those who control it can predict the algorithms of its behaviour.[47,48,49] One of the best analyzes was offered by Lambèr Royakkers et al. humourously asserting that six modern technologies (internet, robots, biometrics, persuasive technology, virtual reality and digital platforms) threaten six areas: privacy, autonomy (decision-making without human involvement – 'out-of-loop'),

---

[40]Moreover, this comparison was already used during the testimony of Frances Haugen, a former high-ranking manager at Facebook, who accused the company in 2021 of knowingly abusing the weaknesses of adolescents.

[41]*Ethics of Artificial Intelligence (2020)*. In: S. Matthew Liao (ed). Oxford Univeristy Press; Dubber, M., Pasquale, F., & Das, S. (2021). *Oxford Handbook of Ethics of AI*. Oxford University Press; Blackman, R. (2022). *Ethical Machines: Your Concise Guide to Totally Unbiased, Transparent, and Respectful AI*. Harvard Business Review Press; Coeckelbergh, M. (2023). *AI Ethics*. MIT Press etc.

[42]Pazzanese, C. (2020, October 20). Great promise but potential for peril. *The Harvard Gazette*. https://news.harvard.edu/gazette/story/2020/10/ethical-concerns-mount-as-ai-takes-bigger-decision-making-role/.

[43]Wachter, S. (2019). Data protection in the age of big dana. *Nature Electronics, 2*, 6–7.

[44]This kind of optimistic comprehensive ethics is offered, for example, by IBM https://www.ibm.com/topics/ai-ethics.

[45]Véliz, C. (2019). Three things digital ethics can learn from medical ethics. *Nature Electronics, 2*, 316–318.

[46]Winfield, A. (2019). Ethical standards in robotics and AI. *Nature Electronics, 2*, 46–48.

[47]Lawton, G., & Wigmore, I. (2023, April 30). AI ethics (AI code of ethics). https://www.techtarget.com/whatis/definition/AI-code-of-ethics.

[48]Giunness, H. (2023, June 16). AI ethics: The ethical issues of artificial intelligence. https://www.techtarget.com/whatis/definition/AI-code-of-ethics.

[49]Bostrom, N., & Yudkowsky, E. (2014). The ethics of artificial intelligence. In: K. Frankish and W. Ramsay (Eds.), *Cambridge Handbook of Artificial Intelligence* (pp. 316–334). Cambridge University Press.

security (hackers who attack everything - from websites to insulin pumps) military drones), human dignity (dehumanisation of healthcare), justice (biometric production of 'impersonators' based on error) and balance of power (dependence on software producers).[50,51,52]

International organizations (UNESCO), states (EU), business and NGO – they all try to define themselves according to the dangers of AI, declarations, recommendations and appeals, which turns out to be rather inept. UNESCO's Recommendation on the Ethics of AI, for example, adopted in November 2021, tries, as is usual with this type of document, to cover everything - the environment, the economy, gender equality, the inclusion of marginalized groups, etc., etc., etc. missing the opportunity for deeper analysis and definition of focus and measures.[53] In this flood of initiatives, there are those who are already concerned about the rights of robots and 'artificial suffering', even those who think that AI should not be used in exchange for nurses, judges, policemen, soldiers and other professions that would threaten human dignity. There are those who believe that artificially intelligent machines have the possibility of destroying human civilization, and those others who bravely reject it (or, as already emphasized, in a few decades we 'progressed' from 'AI is not possible' and 'AI is just automation' to 'AI will solve all problems' and 'AI will kill us'[54,55,56]).

Solutions are sometimes closer than we think. Wouldn't it be simpler to force a factory that lays off workers due to robotization to find a new job for them, giving up the profit generated by robotization (man has to work to feel useful!). Why else would we trade humans for robots, if not to make people's jobs easier, to free them up for more complex or better things?[57]

---

[50]A quantum communications network, developed at the University of Bristol in 2020, raises hopes of overcoming some of the internet's current security flaws.

[51]On the ethical aspects of 'digestible' electronic sensors, see: Gerke, S., Minssen, T., Yu, H. & Cohen, I. G. (2019). Ethical and legal issues of ingestible electronic sensors. *Nature Electronics, 2,* 329–334.

[52]Royakkers, L., Timmer, J., Kool L. & van Est, R. (2018). Societal and ethical issues of digitization. *Ethics and Information Technology, 20,* 127–142.

[53]UNESCO (2023). *Key Facts UNESCO's recommendation on the ethics of Artificial Intelligence,* UNESCO.

[54]Bostrom, N. (2003). Ethical issues in advanced AI. In: I. Smit et al. (Eds). *Cognitive, emotive and ethical aspects of decision making in humans and in Artificial Intelligence* (Vol 2, pp. 12–17). Tecumseh, International Institute of Advanced Studies in Systems Research and Cybernetics.

[55]Eisikovits (2023, July 2023 July 2023). AI is an existential threat – just not the way you think. *BioEdge.* https://bioedge.org/public_health/ai-is-an-existential-threat-just-not-the-way-you-think/.

[56]Stanford Encyclopedia of Philosophy (2023, August, 15). Ethics of AI and robotics https://plato.stanford.edu/entries/ethics-ai/.

[57]Compare with Lufkin, B. (2017, March 7). Why the biggest challenge facing AI is an ethical one. *BBC* https://www.bbc.com/future/article/20170307-the-ethical-challenge-facing-artificial-intelligence.

Adapting (Bio)ethics to Technology and Vice Versa   151

Will AI make mistakes out of ignorance? In his recent novel (Origin, 2017), Dan Brown describes an artificially intelligent assistant who organizes the assassination of a lecturer without malicious intent, to help publicity (what does it mean to 'attract attention?'), to carry out an order. The problem is, therefore, if the machine 'breaks down', so the question is: is the machine a tool in the hands of an evil manufacturer/owner/user or does he generate evil himself (by mistake - that is, he stays with the given procedure even when it brings more harm than good, i.e. he uses ethically/legally illegal procedures)?[58] The next question is: can ethics be programed? If everyone is ethical, how will we be able to recognize/know unethicality?

In July 2023, American authorities opened an investigation into OpenAI for unauthorized data collection and alleged publication of false information about individuals, and a young Egyptian, Patrick Zaki, was also sentenced to three years in prison for spreading false news. A lie is a complex phenomenon: in the past, newspapers used to check what they were writing about, believing that they were doing good (under the influence of ideology or not). So, of course, 'fake news' existed before the era of ChatGPT, it just wasn't prevalent. Now it Machiavellian has become the fundamental motive and the fundamental method of action in politics, trade, education: the lie, devoid of control mechanisms, is today completely integrated into reality, objectified, mimicked, as an uncritical by-product (alternative equal to the truth).

Where Science Is Going: The Play in Which We Decided to Participate

Let's not be fooled: science has long since become something else. Formalization is at work: everything boils down to forms (project applications, project reports, submission of articles for publication in journals, criteria for advancement of scientists[...])

Forms, ostensibly, should contribute to facilitating evaluation, but in fact, they resemble meaningful sabotage that kills originality and specificity. Patterns, of course, unify: but do we want that in science? Algorithms try to replace the human/scientist: patterns are the first step towards the algorithmization of science. The pressure to secure funding brings out the worst in people and results in plagiarism, fabrication and falsification (according to some estimates, 66% of articles published on the topic of COVID are fake).[59] The realization of this is fatal: we fear Chinese, Korean, Japanese, Iranian and Turkish authors, succumbing to the most primitive national and regional stereotypes. Imposing a connection with the economy, commercialization and the profitability of research

---

[58]In 2010, it was proposed to separate the responsibility of the producer from that of the owner/user (*Ethics of artificial intelligence* (2023, September 3). https://en.wikipedia.org/wiki/Ethics_of_artificial_intelligence.).

[59]Carlisle, J. B. (2012). False individual parient data and zombie randomised controlled trials submitted to *Anaesthesia*. *Anaesthesia, 76*(4), 472–479; Ioannidis, J. P. A. (2012). Hundreds of thousands of zombie randomised trials circulate among us. *Anaesthesia, 76(4)*, 444–447.

152 Iva Rinčić and Amir Muzur

leads us to fear speculation: in short, we fear the truly new. We publish in journals that charge for it, that advertise (aggressively), that boast 'open access', that trade in citations (supporting 'mafia cartels'). We think that English is a global language: in most of South America, it means nothing, as well as in the south of Europe, in Eastern Europe, and in the Far East. Shall we finally stand up against specialization, narrowing of knowledge/focus/horizon (Charles P. Snow shows that ignorance is mutual: naturalists know nothing about the humanities, and 'socialists' and 'humanists' about the natural sciences; Van R. Potter emphasizes the same in his 'bridge bioethics'): the result is a flood of works irrelevant to culture. As a countermeasure, in addition to the expansion of education, one could insist on emphasizing ethical reflection (but without referring to a pragmatic 'result', without models and instructions – in the spirit of orientational knowledge promoted by Jürgen Mittelstraß and Ante Čović).

AI Open Questions: Towards an Inconvenient Truth

What (or who) can AI replace/replace? He can read (from) a book in the optimum tone of voice, perfect pacing and diction, up-to-date with the current state of knowledge on the subject: but why would anyone listen to that? It can perform experiments: but will a machine have the idea to run them? Will a machine be able to write a textbook? If so, what will the textbook do for us? AI will question everything, including the meaning of learning. Everything will come down to the question of whether the machine will have Bereitschaftspotential (a conference back in 2009 established that there are machines that can find their own power source, choose the target of attack or develop danger avoidance[60] if we add to the successes of artificial creation the so-called 'xenobots', computer-generated in January 2020 at the University of Vermont from frog cells, with the ability to move and some other characteristics of living beings,[61] it becomes clear that the boundary is not so clear or predictable): if a machine has Bereitschaftspotential, it will become potentially dangerous (like any human). If, on the other hand, he does not have it, he will be a zombie, a servant or a slave: in that case, the danger will come only from his possibility to, without (evil) intention, produce a bad result (false information and harmful move), but then the responsibility will be shared by the machine with the developer/user.

In science and higher education, but also in other industries, the discourse in recent months has come down to the danger posed by ChatGPT and other 'huge language models'. However, the only thing we've found is that those who want to cheat will cheat without ChatGPT; they'll just do it more skillfully now. The problem of ChatGPT (i.e. in a broader sense, AI), therefore, is not in anything

---

[60]Markoff, J. (2009, July 25). Scientists worry machines may outsmart man. *The New York Times*.

[61]Kriegman, S., Blackiston, D., Levin, M., & Bongard, J. (2020). A scalable pipeline for designing reconfigurable organisms. *Proceedings of the National Academy of* Sciences, *117*(4), 1853–1859.

Adapting (Bio)ethics to Technology and Vice Versa **153**

else but in facilitating ethical transgressions that were already known. Fear of ChatGPT is, therefore, fear of ourselves.

References

Note: The reference list is included in a text as a footnote system, as with other footnotes (see main body part).

Allen, C. (2018, February 7). How digitalisation of everything is making us more lonely. *The Conversation*. https://theconversation.com/how-the-digitalisation-of-everything-is-making-us-more-lonely-90870)

Anyoha, R. (2017, July 2). The history of Artificial Intelligence. *Science News*. https://sitn.hms.harvard.edu/flash/2017/history-artificial-intelligence/

Artificial empathy. (2023, June, 17). http://en.wikipedia.org/wiki/Artificial_empathy

Asada, M. (2023, July 21). *Affective developmental robotics: How can we design the development of artificial empathy?*. http://www.macs.hw.ac.uk/~kl360/HRI2014W/submission/S7.pdf

Atkins, D., Uskul, A. K., & Cooper, N. R. (2016). Culture shapes empathic responses to physical and social pain. *Emotion, 16*(5), 587–601.

Bartochowsky, Z., Gatla, S., Khoury, R., Al-Dahhak, R., & Grossberg, G. T. (2018). Empathy changes in neurocognitive disorders: A review. *Annals of Clinical Psychiatry, 30*(3), 200–232.

Baumgaertner, B. & Weiss, A. (2014). Do emotions matter in the ethics of human-robot interaction? Artificial empathy and companion robots. In *Proceedings of the 3rd International Symposium on New Frontiers in Human Robot Interaction*. The Society for the Study of Artificial Intelligence and the Simulation of Behaviour. http://doc.gold.ac.uk/aisb50/AISB50-S19/AISB50-S19-Baumgaertner-paper.pdf

Blackman, R. (2022). *Ethical machines: Your concise guide to totally unbiased, transparent, and respectful AI*. Harvard Business Review Press.

Bostrom, N. (2003). Ethical issues in advanced artificial intelligence. In I. Smit, W. Wallach & G. E. Lasker (Eds.), *Cognitive, emotive and ethical aspects of decision making in humans and in Artificial Intelligence* (Vol 2, pp. 12–17). International Institute of Advanced Studies in Systems Research and Cybernetics.

Bostrom, N., & Yudkowsky, E. (2014). The ethics of artificial intelligence. In K. Frankish & W. Ramsay (Eds.), *Cambridge handbook of Articifial Intelligence* (pp. 316–334). Cambridge University Press.

Briggs, S. (2016, September 12). 6 ways digital media impacts the brain. *InformED*. https://www.opencolleges.edu.au/informed/features/5-ways-digital-media-impacts-brain/. https://www.linkedin.com/pulse/do-robots-viruses-get-sick-days-reflecting-impact-anthony

Carlisle, J. B. (2012). False individual parient data and zombie randomised controlled trials submitted to *Anaesthesia*. *Anaesthesia, 76*(4), 472–479.

Chin, J. H., Haring, K. S., & Kim, P. (2023). Understanding the neural mechanisms of empathy toward robots to shape future applications. *Frontiers in Neurorobotics, 17*. https://www.frontiersin.org/articles/10.3389/fnbot.2023.1145989/full

Coeckelbergh, M. (2023). *AI ethics*. MIT Press.

Decety, J. (2011). The neuroevolution of empathy. *Annals of the New York Academy of Sciences, 1231*(1), 35–45.

Decety, J., Norman, G. J., Berntson, G., & Cacioppo, J. T. (2012). A neurobehavioral evolutionary perspective on the mechanisms underlying empathy. *Progress in Neurobiology, 98*(1), 38–48.

Dimitriadi, Y. (2019). Who you're gonna call? The development of university digital leaders: A case study. *Media Studies, 10*(19), 102–118.

Dubber, M., Pasquale, F., & Das, S. (2021). *Oxford handbook of ethics of AI.* Oxford University Press.

Eisikovits, N. (2023, July 2023). AI is an existential threat – Just not the way you think. *BioEdge.* https://bioedge.org/public_health/ai-is-an-existential-threat-just-not-the-way-you-think/

Ethics of Artificial Intelligence. (2020). In S. M., Liao (Ed.). Oxford Univeristy Press.

Ethics of Artificial Intelligence. (2023, September 3). https://en.wikipedia.org/wiki/Ethics_of_artificial_intelligence

Gerke, S., Minssen, T., Yu, H., & Cohen, I. G. (2019). Ethical and legal issues of ingestible electronic sensors. *Nature Electronics, 2,* 329–334.

Giunness, H. (2023, June 16). AI ethics: The ethical issues of artificial intelligence. https://www.techtarget.com/whatis/definition/AI-code-of-ethics

Gonzalez-Liencres, C., Shamay-Tsoory, S. G., & Brüne, M. (2013). Towards a neuroscience of empathy: Ontogeny, phylogeny, brain mechanisms, context and phsychopathology. *Neuroscience & Biobehavioral Reviews, 37*(8), 1537–1548.

Heyes, C. (2018). Empathy is not in our genes. *Neuroscience & Biobehavioral Reviews, 95,* 499–507.

Hoffman, M. L. (2000). *Empathy and moral development: implications for caring and justice.* Cambridge University Press.

Hren, D., Marušić, M., & Marušić, A. (2011). Regression of moral reasoning during medical education: Combined design study to evaluate the effect of clinical study years. *PLoS One, 6*(3), e17406.

Ioannidis, J. P. A. (2012). Hundreds of thousands of zombie randomised trials circulate amongs us. *Anaesthesia, 76*(4), 444–447.

Kagermann, H. (2015). Change through digitization – Value creation in the age of Industry 4.0. In H. Albach, H. Meffert, A. Pinkwart, & R. Reichwald (Eds.), *Management of permanent change* (pp. 23–45). Springer Fachmedien.

Kriegman, S., Blackiston, D., Levin, M., & Bongard, J. (2020). A scalable pipeline for designing reconfigurable organisms. *Proceedings of the National Academy of Sciences, 117*(4), 1853–1859).

Lawton, G., & Wigmore, I. (2023, April 30). AI ethics (AI code of ethics). https://www.techtarget.com/whatis/definition/AI-code-of-ethics

Lufkin, B. (2017, March 7). Why the biggest challenge facing AI is an ethical one. *BBC.* https://www.bbc.com/future/article/20170307-the-ethical-challenge-facing-artificial-intelligence

Markoff, J. (2009, July 25). Scientists worry machines may outsmart man. *The New York Times.*

Mateš, L., Mladenović, M., & Mladenović, S. (2016). Znaju li studenti prve godine što je internet? [Do first-year students know what the Internet is?]. *Školski Vjsnik: Časopis Za Pedagogijsku Teoriju I Praksu, 65,* 105–117.

Adapting (Bio)ethics to Technology and Vice Versa 155

Matijević, M., & Topolovčan, T. (2017). Izazovi i trendovi u multimedijskoj didaktici [Challenges and trends in multimedia didactics]. *Radovi Zavoda za znanstvenoistraživački i umjetnički rad u Bjelovaru, 11*, 87–99.

Müller, M., & Varga, A. (2019). Digital competencies of teachers and associates at higher educational institutions in the Republic of Croatia. *Informatologia, 52*(1–2), 28–44.

Muzur, A., Rinčić, I., Shim, J., & Byun, S. (2020). Epharmology: A plea for a new science and a new education paradigm. *Nova Prisutnost, 18*(1), 39–46.

Nađ, B. (2020, September 2020). Sve se teže odvajamo od ekrana, ne vodeći računa o zdravlju: Razgovor s Goranom Ivkićem [It's getting harder and harder to separate ourselves from the screen, not taking care of our health: A conversation with Goran Ivkić]. *Universitas*, 20–22.

Neumann, M., Edelhäuser, F., Tauschel, D., Fischer, M. R., Wirtz, M., Woopen, C., Haramati, A., & Scheffer, C. (2011). Empathy decline and its reasons: A systematic review of studies with medical students and residents. *Academic Medicine, 86*(8), 996–1009.

Pazzanese, C. (2020, October 20). *Great promise but potential for peril.* The Harvard Gazette. https://news.harvard.edu/gazette/story/2020/10/ethical-concerns-mount-as-ai-takes-bigger-decision-making-role/

Prensky, M. (2023, May 28). *Digital natives, digital immigrants.* https://www.marcprensky.com/writing/Prensky%20-%20Digital%20Natives,%20Digital%20Immigrants%20-%20Part1.pdf

PricewaterhouseCoopers LLP (Pwc) CEO survey. (2017, December). *Human value in the digital age.* https://www.pwc.nl/en/publicaties/human-value-in-the-digital-age.html

Richtell, M. (2021, January 17). Children's screen time has soared in the pandemic, alarming parents and researchers. *The New York Times.* https://www.nytimes.com/2021/01/16/health/covid-kids-tech-use.html

Royakkers, L., Timmer, J., Kool, L., & van Est, R. (2018). Societal and ethical issues of digitization. *Ethics and Information Technology, 20*, 127–142.

Seppälä, E. (2019, June 11). Empathy is on the decline in this country: A new book describes what we can due to bring it back. *Washington Post.* https://www.washingtonpost.com/lifestyle/2019/06/11/empathy-is-decline-this-country-new-book-describes-what-we-can-do-bring-it-back/)

Sesardić, N. (2022). *Školska knjiga.Konsenzus bez pokrića* [Consensus without coverage].

Smith, C., McGuire, B., Huang, T., & Yang, G. (2006, December). *The history of Artificial Intelligence.* https://courses.cs.washington.edu/courses/csep590/06au/projects/history-ai.pdf

Stanford Encyclopedia of Philosophy. (2023, August 15). Ethics of artificial intelligence and robotics. https://plato.stanford.edu/entries/ethics-ai/

Stueber, K. (2006). *Rediscovering empathy: Agency, folk psychology, and the human sciences.* MIT Press.

Tate, K. (2014, June 3). *History of A. I.: Artificial Intelligence.* https://www.livescience.com/47544-history-of-a-i-artificial-intelligence-infographic.html

UNESCO. (2023). *Key Facts UNESCO's recommendation on the ethics of Artificial Intelligence.* UNESCO.

Véliz, C. (2019). Three things digital ethics can learn from medical ethics. *Nature Electronics, 2,* 316–318.

Wachter, S. (2019). Data protection in the age of big dana. *Nature Electronics, 2,* 6–7.

Weisz, E., & Zaki, J. (2018). Motivated empathy: A social neuroscience perspective. *Current Opinion in Psychology, 24,* 67–71.

Winfield, A. (2019). Ethical standards in robotics and AI. *Nature Electronics, 2,* 46–48.

Yalçin, O. N., & DiPaola, S. (2020). Modeling empathy: Building a link beween affective and cognitive processes. *Artificial Intelligence Review, 53*(4), 2983–3006.

Yamamoto, Y., & Ananou, S. (2015). Humanity in the digital age: Cognitive, social, emotional, and ethical implications. *Contemporary Educational Technology, 6*(1), 1–18.

Zovko, V. (2016). ICT-enabled education – Need for paragidm shift. *Croatian Journal of Education, 18*(2), 145–155.

Section 3

Space

Chapter 8

Are Space Technologies Untimely?

Tony Milligan

King's College London, UK

Abstract

Suspicions about space technologies can be regarded as instances of "space skepticism," i.e. the broadly pessimistic view that human activities in space are untimely or liable to be counterproductive. Section 1 will explain that contemporary space skepticism is focused upon negative societal role rather than the physical possibility of the activities proposed. Such skepticism is a complex pool of familiar claims unevenly drawn upon rather than a single theory. Section 2 will suggest that contemporary space skepticism tends to integrate with a broader set of doubts and fears about technologies of the Anthropocene. Section 3 will draw out a tension within the skeptical complex between the idea that concern for space is irrelevant to our societal problems and the idea that it is likely to make such problems worse. Section 4 will briefly outline why the publicly dominant forms of space skepticism carry a growing capability for merger with political activism and why the publicly dominant skepticisms are not necessarily those with the greatest plausibility, but rather those with the strongest motivational force. Finally, Section 5 will set aside the issue of popular motivational force and focus instead upon the skepticisms with the greatest plausibility. It will briefly outline why the relevant fears and suspicions (about military tensions and geoengineering) are outweighed by other considerations. This response to plausible skepticisms will not aim to be comprehensive but indicative of the direction of travel for more detailed critique. It will provide a framing context for a large metaphor about space technologies *allowing the Earth to breathe*.

Keywords: Space skepticism; timeliness; geoengineering; war; Anthropocene; political activism

---

The Ethics Gap in the Engineering of the Future, 159–175
Copyright © 2025 Tony Milligan
Published under exclusive licence by Emerald Publishing Limited
doi:10.1108/978-1-83797-635-520241009

160   Tony Milligan

Setting up the Problem

Skepticism about human activities in space comes in a variety of forms. The oldest versions involve claims of impossibility. So, for example, Robert Goddard's early work on rocketry in the 1920s was ridiculed on the grounds that rockets would have nothing to push against in the vacuum of space (Smith, 2018). Following the launch of Sputnik in the 1950s, a more modest skepticism allowed that rockets would work in space, but that human bodies would fail if pushed through the Van Allen radiation belts on the way to Moon. The latter claim is still cited by conspiracy theorists about the Moon landings (NASA, 2010). But what interests me here is space skepticism (Milligan, 2015) in the sense of a generalized hostility towards or doubt about the value of space programs, space science and human activity in space. For convenience, conspiracy theories form an unusual and special class of skepticisms which will be set aside because they raise a different series of issues about standards of evidence and about what separates out science from pseudo-science and reliable testimony from the sensationalist and the absurd. My concern will be closer to the mainstream of doubt about the value of human activity above the Kármán Line and with a set of ideas about value rather than physical possibility. From the mid-20th century, with the reality of space travel accepted by almost everyone, skepticisms have tended to move away from claims about physically insurmountable difficulty towards more societal and value-based concerns.

As an exemplar of this shift, Amitai Etzione's *Moondoggle* (Etzioni 1964) highlighted the risks of a brain drain from other projects, the apparently high taxation costs and the apparently luxury status of space expenditure. The latter concern resurfaces in Arnold Toynbee's seminal *Mankind and Mother Earth* (1976) where space exploration is compared to the building of the pyramids (Toynbee, 1976). A comparison which works both ways: yes, it was costly, but do we really want to live in a world without pyramids? By this time, Malcolm X had already criticized those sections of the black community in America who had been seduced by the space programme, at the expense of more burning social concerns (Malcolm X 1963), and this theme had set a baseline for critiques associated with the late Civil Rights movement, leading to a well-known protest at the gates at Cape Canaveral on the night before the launch of Apollo 11 (De Groot, 2006).

Echoes of the focus upon cost and waste continue in contemporary doubts about the value of human activities in space. However, there is something odd about trying to make sense of a civilization level change primarily in terms of what turn out to be modest alterations in taxation. Similarly, with the rolling out of robotization over the remainder of the 21st century. Rather than being an expense within a relatively static economic system, the likelihood is that the economic system will itself be transformed by robotization in ways that will make familiar 20th century arguments between fiscal conservatives and liberals or social democrats outdated. Something similar can be said about the space economy. Unless it accounts for a large portion of taxation, the latter can plausibly be regarded as the wrong metric to use.

Are Space Technologies Untimely? **161**

This rather prosaic discussion about cost has been accompanied by attempts to frame matters in similarly skeptical but deeper and far more civilizational terms. Indeed, the deeper discourse preceded the taxation concern. Lewis Mumford's *The Transformations of Man* (1957) exemplifies this more thoughtful level of concern, with its worry about an outward turn to space taking primacy over an inward turn to humanity. A view which some clear precedents in religious concerns about modernity and technology shifting our attention to the outwards and the surface at the expense of our humanity and spirituality. (Concerns which restructure familiar 19th century religious critiques of capitalism as a great underminer of our proper inward focus.) Hannah Arendt's classic essay "The Conquest of Space and the Stature of Man" (1963) also sits somewhere in this territory with its unease about the view of humanity from above, as little more than a collection of rats running around in mazes, convinced that they are in charge of processes when action and direction are shaped by something else (Arendt, 1963). Such skeptical, and value-based worries did not end with the arrival of the Apollo programme. J. G. Ballard's fictional Cape Stories, written from the Gemini programme onwards through both the Apollo and the Shuttle eras, continued this line of spiritualized and skeptical attack. For Ballard, while spaceflight drew upon a natural human urge towards transcendence, it pursued a false transcendence of the human. We needed a different kind of transcendence from the one sought by Dante and the traditions of Western religion and art, but spaceflight and the premature space age had attempted to move on from the latter in the wrong way, through *fugue*, through escape from our mortality (Ballard, 1981). All of which is more philosophically deep than debates about taxation.

Skepticism of this sort has always tended to accept that humans will eventually go into space, in the natural course of events, but only when the time is right and without political or technological forcing. This is a much harder worry to address. It seems perfectly reasonable to point out that it would *not* have been a good thing for our emerging space technologies (not just rockets but even satellite systems) to have been available at various points in our recent human past. It is good that such technologies were not available in the 1930s, ready to hand for use by the dangerous but technologically savvy regimes then swallowing up Europe. We might also argue about the advisability of the availability of ICBMs at the height of the Cold War when the rudimentary nature of the technologies offered little reason to be confident about the ability of either side to win in any first strike scenario. In a thought experiment where we are transported back in time, without knowledge of outcomes, and with the option of pushing a button to rid the world of the dual-use (military and non-military) rockets in question, it would have been tempting to push the button. Dangerous technologies, and dual-use technologies in particular, are not always timely.

This issue of timeliness has gradually moved closer to the centre of skeptical concerns. Even when there is a reversion to public arguments about taxation and expenditure, a concern about timeliness generally remains part of the mix given that few people want to take away the possibility of a future in which clean technologies and easy movement between Earth and the rest of space is available. At some point in time, most people want this to happen or believe that it will

162 Tony Milligan

happen and that there is nothing particularly wrong about it happening as long as it is *then* and not *now*. The guiding thought of space skepticism, when temporalized in this way, is that we should not be spending so much on space *at a time like this*, or *while societal problems remain unsolved*, or *while we still have not fixed climate change*. An exemplar of this is the best current statement of space skepticism, i.e. Daniel Deudney's *Dark Skies: Space Expansionism, Planetary Geopolitics, and the Ends of Humanity* (2020) "The pursuit of ambitious space expansion must now prudentially be judged to be deeply undesirable for humanity and the Earth for at least several centuries" (Deudney, 2020, p. 381).

Given that technologies are usually untimely with respect to something, there is no knock down argument against such skepticism. Modern computing has brought a golden age for stalkers, the mobile phone has led to technologically enhanced bullying and one of television's first notable roles was as a vehicle for Nazi propaganda. There are upsides and downsides of socially transformative technology and the two do not come easily apart. As a result, a more fine-grained framing of what I will call *the problem of timeliness* is appropriate. A framing that does not concern untimeliness of any sort whatsoever but is instead restricted to *untimeliness relative to the major challenges currently facing humanity*. Most obviously, the challenges of climate change, conflict mitigation and socially corrosive forms of injustice.

With respect to these matters, space technologies do raise concerns although certain kinds of concern about societal injustice may have little purchase, given that diversity and inclusion are strongly driven within the space sector. As we might expect with economic activity which lacks the historic inertia of heavy industry with its heavy composition of white males within the traditional blue-collar workforce. Indeed, there is a case for saying that a sole focus upon diversity and inclusion might lead us to favour a more rapid expansion into space, as a mechanism for levelling up, rather than more of a slow space approach which can offer a greater likelihood of environmental protection in remote places (Milligan, 2023a). Accordingly, I will focus less upon how space technologies play out in terms of the problems of diversity and inclusion and more upon the interlocking of space technologies with military tensions and climate change response.

The Targeting of Technologies of the Anthropocene

Space technologies such as satellites and communications systems, rocketry and propulsion systems, as well as habitat infrastructure do tend to be dual use by design and not by accident. They typically have military as well as civil applications. As an extreme case, think of the propulsion technology required for any sort of journey to the stars or, more specifically, towards Proxima Centauri. The speed (and energy) required for any space vehicle could easily make it into the most dangerous bomb in history if turned in the other direction, i.e. back towards Earth. Rather than extending humanity's survival chances, it could help to compromise them. Less dramatically, and back on the ground of actually existing (and currently emerging) technologies, even technologies (space harpoons and all sorts of nets) developed as ways to deal with the growing problem of space junk in low Earth orbit (LEO), and to some degree also in geosynchronous orbit (GEO), could potentially be used against non-junk. Or they could be used for military intelligence gathering in the case of junk which is not actually rubbish but tech which has simply been decommissioned or has fallen out of use.

Here we might also reflect that Russia can no longer level-up with the technological development of the US and its allies, but it could try to level down the playing field by taking out the space technologies of others. A Gordian Knot solution to the growing asymmetry in military capability. Damaging tech often tends to be easier than duplicating it. In this context, it is worth remembering that the militarization of space has always been an aspect of our human activity space. A strict pacifist would probably have difficulty supporting any actual space programme because this knot, tying military and space programs, cannot itself be untangled. NATO formally declared space a sphere of operations in 2019, and a series of space forces are currently being formed by the relevant nations. While there is a widespread notion that militarization per se is illegal under international law, this is also one of the more familiar misconceptions about space law. The Outer Space Treaty (1967) is geared towards containment of military operations in space, rather than their exclusion. It only excludes nuclear weapons and weapons of mass destruction. Not space weapons per se (Coleman, 2022). One reason for this is the rough and ready pragmatism required to get agreement on the Treaty. Another, is the fact that exclusion would also prevent almost all civil activity in space given some unavoidable level of dual use, as well as dual use by design.

Space expansion and some degree of militarization do go hand in hand. Yet militarization and actual conflict are not one and the same. Their relation is complex rather than linear. Deterrence can unintentionally lead to war. Military build-up can lead to unexpected peace. Tendencies and counter tendencies play off one another in complex, situation-dependent ways which are often clear only in retrospect. On paper, and with no other information, the colossal weapons build-up from the 1950s onwards had global war written all over it. That could have happened. At times it seemed likely. Yet it did not take place and the balance sheet for the impact of the weapons build up is (arguably) positive. Peace was largely maintained, with regional exceptions and a balance of military power was one of the reasons why it was largely maintained.

The most striking concern about militarization in the case of space, and a reason why we cannot simply project a similar preservation of peace into the future, is that the stakes are higher than during the Cold War. The use of nuclear weaponry in the 1960s could have resulted in untold suffering and system collapse, but it is unlikely that humanity would have been entirely wiped out. Space weaponry raises the stakes to extinction levels. Deudney has been quick to point this out. Asteroid redirect capability to protect the Earth is also the capability to use kinetic weapons directed against it (Deudney, 2020). The point is much the same as the one made above that interstellar vehicles can function as relativistic weapons because of their enormous speeds. But it is much more worrying, given that we will soon have asteroid redirect capability, whereas interstellar craft will remain fictional for some time. (Perhaps always.) At some not too distant point in time, we will need to be able to redirect and deflect asteroids, and this will come at the price of weaponization by rogue states and agents. It does make a good deal of sense to ask whether or not we should take the risk of having this kind of dangerous technology. In principle, a thoughtful skepticism which runs along these lines could be persuasive. It could help to convince us that the net result of asteroid capture technologies is likely to be negative and that such things will be a menace to humanity. A skepticism of this sort could also escape from the change of technophobia, given that it is specifically about a genuinely threatening technology and not a blanket response to advanced technologies of any sort.

One of the interesting things about this line of thought is that it broadens the grounds for space skepticism, or rather for space skepticisms, i.e. for holding that human activity in space will be of negative value. So far, I have framed such skepticism in terms of "temporalization": the idea that now is not the time. But fears about some ultimate catastrophe, enabled or driven by technologies such as interstellar craft or asteroid capture tech, may instead be framed in terms of a "teleological claim": for one reason or another, space expansion will not just end badly but has an inbuild drive to end badly. The temporalization of space skepticism, and this sort of teleological claim both help to support an emerging association of space skepticism with broader forms of social critique, with the idea that a political forcing of space activity by elites, when we are not ready for space, will end badly for all concerned. For humanity as a whole.

Arguments along these lines involve elements of "populism" or environmentally inflected populism. The one percent are branded as Earth abandoners while the rest of us, the 99%, must turn towards Earth. Populism of this sort also draws upon two further core claims. One concerns "attention," the idea that ambitious space activity diverts concern and resources away from Earth. The other concerns "identity": a focus upon space cuts against our recently achieved sense of being Earthlings or belonging to Earth. Within the academy, such ideas draw upon the down to Earth approach of Bruno Latour (2017), albeit selectively (given that Latour has no problem with seeing the Moon and Earth as a single system across which human activity will become increasingly normal).

In some respects, the identity claim echoes Hannah Arendt's concerns about how activity in space will reshape the ways in which humanity sees itself. However, Arendt's concerns were of a more philosophical sort than the populist critique of Earth abandoners. Arendt's writings were never likely to result in protests at the gates of launch sites. An identity-inflected skepticism geared to environmental dissent and elite hostility is far more likely to do so. There have been only occasional instances of any connection between space skepticism and dissent over the past half century. But a changed and closer relationship is emerging, exemplified by a satirical "1%" video released in 2021 by Greta Thunberg and *Fridays for Future*. A mock advert encouraging the elite to move to Mars, a pristine planet, now that they have wrecked the Earth. The rest of us, the 99%, had better do something to fix climate change (Milligan, 2023b). The picture

Are Space Technologies Untimely? **165**

painted is one of friends and enemies. Battle lines are drawn, and the enemies are both cohesive and focused upon space. This is 21st century populism, and far closer to Latour and Carl Schmitt than to Karl Marx.

It is also noteworthy that there is an equivocal pattern in the associated activism. A targeting of emerging technologies in ways which allow for plausible denial that they are the target. In 2021, Thunberg was involved in successful protests alongside Sami about the conducting of rudimentary and non-intrusive geoengineering experiments in Sweden, i.e., high altitude balloon tests, linked to ongoing geoengineering research at Harvard and supported by the National Academy of Science (Dunleavy, 2021; Wibeck et al., 2015). The Swedish space agency called off the test. The protest itself was not against actual geoengineering, but against research geared to evaluate its practicality at some later point in time. In 2023, the emerging relationship led to protests over the continuing presence of wind farms on Indigenous Sami lands in Sweden, in defiance of a court decision for their removal (IWGIA, 2023). A cause with which we might sympathise on grounds of support for Indigenous rights, yet the recurrence of a technological focus is striking. These protests have a special salience given that Sweden has one of the most important European satellite launch sites, Esrange near to Kiruna, on lands that Sami reindeer herders have traditionally ranged across. The protests occurred in the shadow of emerging space technologies.

In multiple locations across the globe, there has also been an entanglement of space ground infrastructure with issues concerning Indigenous rights, and this can feed a perception of dissent linked to space skepticism as socially just, and space technologies as part of the colonial machinery. The main launch site for the European Space Agency is in French Guyana at a location entangled with colonial legacies which run against the image of progress and the inclusive future that ESA wish to cultivate (Redfield, 1996, 2000). A reminder that futures cannot wholly disentangle from the past. But none of this means that there is precisely one globalized Indigenous attitude towards space, or that such an imagined single attitude aligns with a 99% set against a cohesive 1% space enthusiast elite. Similarly, we might sympathize with worries about geoengineering and about land use while being concerned about a repeated tendency of populist dissent to target technologies when there is no shortage of other environmentally linked injustices that might be targeted.

For convenience, I will refer to these as technologies of the Anthropocene, by which I mean technologies which have emerged to enable human activity in space, together with robotics, AI and bioengineering (The list is indicative rather than exhaustive.). The technologies in question are those which back up the idea that our human impact upon Earth now runs geologically deep and is getting deeper all the time. Concern about technologies of this sort, associated with global human impact, no doubt draws upon background suspicions about technology per se, but it also draws upon an idea of moral hazard. An idea that technological readiness insures against bad behaviour and can therefore encourage it. Such readiness, or even a focus upon technological response, distracts from the pressing need for personal, individual change. If we say that various technologies can become important parts of climate change response, then we may encourage the idea that technology will come to our rescue, that we can afford to behave badly. Within this picture, space technologies are just one of several guilty, or super-wealth-related technologies of the Anthropocene which bring us false hope that something other than dissent and changed personal lifestyle can fix climate change. There may be a large problem with the associated understanding of how behaviours change at a societal level, in some way that is independent of technological change. But there is an even larger problem of imagining that behavioural change alone can constitute an adequate societal response. Even with best human behaviour, the Greenland icesheet will still disappear. Technological response has to be some part of the mix.

In a sense, we might even consider that humanity has been extremely fortunate in point of timing. Had space-based Earth monitoring technologies emerged any later, then we might still be utterly blind as to what is happening. Moral hazard suspicions about technologies of the Anthropocene have little purchase in the case of space technologies when the very identification of a climate change problem at a global level has been, from the outset, utterly dependent upon these technologies. Of course, not every space technology offers obvious advantages for climate response. Those associated with space tourism are far more tenuously connected.

This consideration does figure in the targeting process for dissent. The 2021 *Fridays for Future* video was released in the midst of a race into sub-orbital space by the leading space tourism companies, Blue Origin (associated with Jeff Bezos) and Virgin Galactic (associated with Richard Branson). And such activities, strongly connected to the idea of billionaires in space, form something of an easy target for populist inclined critique. The optics are not always good. Moreover, their standing as extensions of more conventional commercial flight makes them a tempting target for environmental movements which have always been shy about promoting the less popular messages that emissions from road transport and meat production utterly marginalize the emissions contribution from flight. Flight and experimentation are softer targets than diet and the family car.

Tensions Within the Skeptical Complex

The above provides an outline of what I take to be the prevailing skeptical complex and an indication of some of its main drivers. These drivers include familiar suspicions about technology, and particularly about advanced technologies of the Anthropocene. And they include the fear that a need for behavioural change will be lost if we focus upon technological response. Space technologies will always, then, seem untimely when there are pressing Earthly problems that they draw attention away from.

I have also suggested that there are some internal tensions within the associated skeptical complex. One in particular, stands out: a tension between the idea of misplaced attention and the teleological claim. At this point, a little unpacking of the teleological claim will be useful. Generally, the idea is that space expansion will generate new problems and/or regenerate older problems. Again, typically those of colonialism although it would be extremely difficult to reproduce anything like an actual colonial system, or a true colonial mentality, in the absence of any subject Indigenous peoples of space. The worry is more about a mentality of domination and conquest, and occasionally, more science fiction (specifically cyberpunk) worries about the creation of a disadvantaged space proletariat. An unlikely outcome given the large skill set required to survive in space and the prospect that much of what is done in space will be robotic. It is far more likely that we will generate new socio-economic problems stretching our sense that the space economy has any strong connection to the world of 19th century mining or the textile mills of early capitalism. Our dominant socio-economic systems are not now in their infancy. The more plausible versions of the teleological claim (that space is driving us in a bad direction) involve the idea of a new and technologically mediated set of problems.

As a high-level description of such problems, they involve negative feedback in the form of the pressures towards geoengineering which have already been noted, and increased military tensions generated by considerations of strategic advantage and the limited resources of accessible space. Competition for limited resources can look superficially similar to the militarized competition of the classic colonial era. And talk about terraforming Mars might help to normalize geoengineering on Earth. With these aspects of the teleological claim unpacked, the tension with the idea of misplaced attention becomes clear. Misplaced attention concerns drawing away our focus from terrestrial problems. It rests upon an idea that space is in some way irrelevant to our most pressing concerns. A diversion. Space activities and the related technologies are not going to impact upon what we really care about so why spend so much time and money upon them? By contrast, the teleological claim is precisely that space activities and the related technologies *are* going to impact upon Earth. They are going to impact in a large and significant way, and their impact will be negative.

This does amount to a tension. Tensions can, however, be normal across sets of claims and so we should not take their presence as an automatic *reductio* or deconstruction of what opponents of space expansion are saying, and saying *in the round*, i.e., in the light of a variety of considerations. It may, however, lead us to ask which of the claims captures the deepest level of concern, and which of the claims pose the greatest challenge to those who hold that space technologies and space expansion are timely. Here, I will suggest that the deeper, more plausible, and more informed claim out of these two is the teleological claim rather than the claim of misplaced attention. This is not to say that I agree with the teleological claim. Indeed, I think that we are poorly placed to determined what the telos of space expansion might happen to be. Matters could go well or badly. There are some plausible reasons to hold that there will be harms but the overall outcome may be beneficial at a civilizational level, and that space expansion and space technologies are not some pathways to dystopia. However, we are epistemically disadvantaged when it comes to determining the overall impact of radically transformational technologies and tend to fall back upon reasonably informed guesswork, social hope and an awareness of how things might go in the light of the recent past. We might find reflections of this sort reassuring, but they are not guarantees.

168   Tony Milligan

My point that the teleological claim is the more significant, more plausible and deeper claim should not be taken as actual endorsement. Rather it is a judgement about relative merits, and about the teleological claim's advantages when contrasted with a misdirected attention claim which has tended to be more intuitively appealing in critical public discourse but which would be difficult to support on any deeper dive into the actual societal role of space programs. As a further clarification, I am *not* suggesting that this is how sceptics themselves must see matters. Indeed, there is some case for saying precisely the opposite, i.e., that the dominant form of space skepticism now plays far more upon the idea that space is irrelevant to our human predicament, rather than the idea that it is shaping our terrestrial responses in ways that will end badly. The "1%" mock advert exemplifies this preference for the weaker line of argument, by playing upon escapism rather than a deeper set of concerns about a potentially negative impact of space activities upon the global balance of power and associated military tensions. The latter involves a less streamlined set of concerns.

The Partial Merger of Space Skepticism With Political Activism

Another way of making the above point would be to say that space skeptical activists are responding to a different set of problems associated with political mobilization rather than intellectual justification. Historically, it has been difficult for any form of space skepticism to merge with activism, although there have been periodic attempts to make this move. (The Malcolm X speech and the Apollo 11 protest are obvious examples.) The populist component of the new skeptical complex is making such a merger with political activism more likely, and this populist component draws heavily upon the streamlined idea of irrelevance to present and immediate planetary problems. More heavily than upon the teleological claim about future misfortune. It is also part of the space skepticism complex, but populism leans towards the here and now. Simple and immediate solutions for difficult problems. And while we might wish that activism was uniformly grounded in strong forms of rational justification, the relationship between such justification and activism in general has always been more complex. Partly because our human motivations and justifications are not always the same, and reasons for action which are offered can be little more than rationalizations of the things that really motivate us to act. This can also raise ethical concerns about the influence of racism, antisemitism and puritanism upon patterns of dissent.

Another way of making the point would be to imagine the big bowl of possible activisms, figuratively competing for the attention of potential activists who can only choose so many from the bowl. Some are eye catching, others less so. Some win out, others loose, but there is a pattern to this winning and losing and it has very little to do with what is right and wrong. Instead, the winning activisms lean into *what will motivate*, rather than leaning into *the strongest form of rational justification*. These two may fortuitously coincide, but they need not do so. The motivations themselves can be bad, indifferent, extremely good, but they are often unexamined, or else the justifications offered are themselves mistaken for motivations and so no further enquiry is considered necessary. As a qualification, activisms are never indifferent to justification, and it usually figures prominently as an aspect of theory building among activists themselves, creating something of a split between internal discussions and public engagements. Yet the separation is never absolute and the relation between the two is complex. And so, while activisms often favour simple solutions or one solution, activisms themselves are complexly structured.

In the past, space skepticism has not done particularly well in the competition for activist attention. It has been present, but generally left in the bowl. A streamlined connection to the idea of Earth abandonment by a technology loving elite makes it much more attractive from a motivational point of view for a significant number of activists. This is an activism that will be selected, especially by activists who are generally suspicious about technologies of the Anthropocene, suspicious about who might control such technologies, and comfortable with a focus upon making the world better through a combination of protest and personal behavioural transformation.

However, the prospect of a closer connection between political activism and space skepticism does raise some large ethical concerns from the standpoint of the ethics of political dissent and risk management at launch sites. Especially if we regard the avoidance of reckless endangerment as an ethically significant aspect of determining whether or not particular instances of dissent are justified, and the standing that they might have, e.g. whether or not they might count as civil disobedience. Launch sites tend to be in relatively isolated places, and newer launch sites tend to be more secure than older sites. However, this is unlikely to avoid Greenpeace style activism geared to accessing difficult locations such as oil rigs in the Norwegian Arctic. The prospects for such activism may pose strategic challenges when it comes to siting decisions for ground-based infrastructure. Having a large body of protestors at the gates or behind a low wall only hundreds of metres away when a liquid fuel rocket might go badly wrong, would not be an ideal option, yet there are sites which are like this. The Xichang launch site in South-West China comes to mind, and its over exposure seems to have influenced the configuration and ground clearance of the newer island site at Wenchang which is taking over its functions (China Space Report, 2024). In spite of the activism-related advantages of the more streamlined and populist inflected misdirected attention claim, there are strong ethical reasons for caution about any ensuing activisms. Having any group of people trying to get to a launch pad when unaware of safety concerns and the risks would be problematic.

And, if we revert simply to considerations of justification, the misdirected attention claim does not do at all well. An appeal to the teleological claim and the idea that space activities are misdirecting us towards a bad end, has at least the advantage of appealing to the uncertainty of the future, rather than having to explain away the ongoing environmental advantages of planetary knowledge that human activity in space has made available. Most obviously, through satellite technologies used to identify, track and respond to climate change. More qualified versions of space skepticism do, however, tend not to target every kind of activity in space, but rather larger ambitions beyond LEO and GEO as well as sub-orbital

170   Tony Milligan

flights (space tourism) as a frivolous elite activity. However, this is problematic as well as (in the latter case) drawing upon something that looks suspiciously like a form of puritanism. It is problematic because certain kinds of broadly "Western" societal backgrounds do seem to correlate with the experience of an "overview effect" (White, 2023) or transformational "orbital perspective" (Ron Garan, 2015) in space. And this is something that is surely a good thing. There is a case for saying that economic decision-makers, and perhaps politicians *should* routinely be sent into space, to gain this larger perspective of Earth as the integrated and valuable place that they are in office to protect and serve. If every COP meeting of the UN to set targets for carbon emissions was preceded by a significant number of delegates going sub-orbital then outcomes might well be more ambitious. Figuratively, this is the environmental equivalent of suggesting that slaughterhouses might have glass walls as a way to encourage non-meat diets. We might worry about the associated flight emissions, but they would be marginal compared to those of such events as a whole, just as emissions from flight are generally negligible compared to those routinely associated with road traffic and meat consumption.

And, if we think instead about activities beyond LEO and GEO, the idea of space activities as irrelevant to terrestrial (especially environmental) concerns fails to get off the ground. Rapid identification of a greenhouse effect on Earth when satellite data of a sufficient volume and quality began to come in during the 1980s was heavily and directly shaped by the atmospheric modelling of Mars and especially of Venus, where a greenhouse effect had already been identified (Leishner & Hogan, 2019). More precisely, the very idea of a greenhouse effect emerged from the modelling of Venus. And our understanding of the systematicity of Earth's climate was already strongly shaped by both the Mars and Venus research. As a symptomatic point, James Lovelock, who pioneered the Gaia hypothesis so beloved of terrestrial environmental movements, was a Mars-Venus atmosphere researcher with NASA during the 1960s. It was out of this work, and the comparison of atmospheric systems and ground on Mars, Earth and Venus that Gaia emerged (Lovelock, 1965).

Of course, we might set aside ongoing satellite monitoring, look at broader forms of planetary research elsewhere and say "That was then, and this is now. What have space programs done for us lately?" But here, I would point to our pressing need to understand how polar ice behaves on other planets, and how ocean systems behave on other water worlds, given that Earth too is a water world with polar cryosystems. Planetary science is simply better when it is based upon more than one example, just as our understanding of games is better when based upon more than an understanding of one single game (no matter how good the game may happen to be). And while it is true that a great deal of the prospective activity in space, such as asteroid mining, will not be directly scientific, it is also true that scientific research is not free floating. An expanding system of scientific research, with bases in strategic locations (such as the south lunar pole, both Martian poles, and in orbit around water worlds) requires a much broader system of support and infrastructure. On Earth, comparable systems are both commercial and military, and the same can be said about aid programs although it is politically inexpedient to draw too much attention to the background commercial and military systems of support.

These points do not strike home in the same way when we consider the teleological claim, the point of which is precisely that human activities in space are *not* an irrelevance, they *do* carry impact but the impact is going to be negative and negative in a large way. Where this leaves us is with an understanding of why a weaker skeptical line of argument is often preferred over a far stronger but equally skeptical line of argument to the effect that space activities and the associated technologies are driving humanity in the wrong direction. This stronger line of argument plays well off the point that there is no absolute guarantee that the future role of space technologies will be strictly benevolent. There have already been injustices associated with ground-based infrastructure and we know that key technologies do often (perhaps always) come with downsides. They are untimely with respect to something. Recognition of this can and should feed into any bottom-line assessment of whether such technologies are timely or untimely *overall* or *with respect to the main challenges facing humanity.*

Timeliness and Risk

Above, I have argued that the most plausible version of space skepticism concerns the future that space technologies are driving us towards. I will expand upon this a little and conclude with an overall assessment of how much weight we should give to the genuine problems that this form of skepticism identifies. Among these, two stand out from the others: space activity is bringing a future of heightened military tensions and dangerous technologies, raising the prospects of war and further damage to the planet through misguided attempts at geoengineering. There is no great step from (currently impractical) talk about terraforming Mars, to slightly more practical talk about terraforming Earth, in the sense of moving towards a geoengineering paradigm as our final line of defence in the face of climate change.

Both of these future oriented concerns are serious, identify serious risks and can draw upon at least some evidence. By contrast, skepticism based upon claims about misdirected attention is counterevidential. In the case of geoengineering, we can look to the considerable environmental damage and injustice caused by the great dam building mega projects of the 20th century. Damage and injustice that we are, in some cases, currently trying to undo (Wildman, 2013). In the case of military tensions, we can look back to the era of ICBMs and to the integration of the first space race into the Cold War *as an aspect of the latter* rather than an external outcome of it. Within the space community it is familiar to point to space programs as driven by Cold War tensions and elide over the point that they were part of such tensions and integral to them. Like it or not, human activity in space has always been at least partly militarised. This historic trend *is* likely to continue, and it does raise the prospect of conflict at some point. Not in the form of great battles in space that we see in *Star Wars* or *Battlestar Galactica*, but in less dramatic forms such as the use of electromagnetic pulses to disable satellites, the directing of technologies for space junk against non-junk, the deliberate creation

172 Tony Milligan

of debris fields rendering LEO unusable, and perhaps even the use of asteroid redirect technologies to throw large metallic objects at the planet.

The greater of these concerns, military tensions and geoengineering, is probably the former. Geoengineering could cause great damage, but it is unlikely to pose an extinction-level or civilization-collapsing threat, even if it does go wrong. If it is attempted, and does go wrong, then we will probably stop and live with the resulting damage. A high degree of caution may be anticipated. By contrast, it is difficult to imagine the use of an Earth redirected asteroid as a kinetic weapon proceeding with caution or being anything other than utterly catastrophic for humanity. The larger, and pressing, concern is about the increasing militarization of space and where this might ultimately lead.

When deliberating about this, and the rising military stakes in space, it is worth repeating the point that the capability to destroy space infrastructure may be less technologically demanding than attempts to match it. Russia is no longer in a position to compete technologically with the USA, so its best strategic option may be to level-down the playing field by reducing US capability. (Through direct damage to satellites and/or the creation of debris fields.) Relatedly, the intractable problem of rogue states and apocalyptic terrorist movements raises the alarming prospect that a balance of power between major military players may not be enough to secure the Earth from threats. Space technology is increasingly dual use by design rather than by accident, raising the prospect of dangerous kinds of misuse driven by ideologies which do not necessarily promise the survival of those who initiate action. We cannot presume that rational self-interest will safeguard us from the most destructive behaviours. We know enough about rogue states and terrorist organizations to realise that suicidal action would not be an absolute deterrent.

Well-informed skeptics, such as Daniel Deudney in *Dark Skies*, have pointed towards the risks. However, I will suggest that the deliberate creation of debris fields, rendering LEO unusable, may be a more likely scenario than turning asteroids into civilization-threatening kinetic weapons. The technology required to redirect asteroids is much the same as the technology required to protect Earth from asteroids, and given our distance from the main asteroid belt, there would be considerable time to deploy protective measures. This is certainly not an absolute safeguard, given that any and all protective measures could (in principle) be taken-out. But even those like Deudney, who draw our attention to the problem, and call for the renunciation of our larger space ambitions for several centuries, tend to accept that we should be developing at least some protective measures for asteroid interception because the threat in question will remain a threat even in the absence of malevolent agents. At some point the Earth *will* be hit if we do not take steps to prevent this from happening.

This might not be for a very long time, but it seems unwise to rely upon good odds in order to defer the development of planetary defence. Doing nothing because the odds are currently in our favour would be a bit like playing Russian roulette with a gun that has a vast number of chambers. Some of us might feel that a sufficiently large number of chambers, say a few thousand, would remove any great personal fear. But we would hardly force others into such a game, or

Are Space Technologies Untimely? **173**

(figuratively) point it at the head of humanity as a whole. It would be wrong to do so irrespective of the number of chambers that might be added. The analogy is, admittedly, imperfect because Russian roulette involves an act rather than an omission (doing nothing), but for those who are in the business of worrying about civilization-level threats, the prospect of an asteroid strike may be sufficiently alarming and realistic for the analogy to do enough work to suggest that indefinite delay is not a good option. And this is Deudney's approach. However, once we accept the need for planetary defence, we are already in the territory where risks also kick in. We cannot have such defence and guarantees against the malevolent use of asteroid redirect technologies.

It is also unclear that there will ever be a politically ideal time for us to develop the technologies associated with such a risk. The idea of liberal democracy gradually extending to cover the globe is currently difficult to sustain, and for the coming centuries our best hopes may be for some set of hybrid systems, democratic up to a point, holding together a balance of power while humanity makes its way through the bottleneck of a period of risk associated with space technologies and technologies of other sorts. For this purpose, a sufficiently large presence of liberal democracies committed to a stabilization of international relations and the maintenance of a balance of power may be non-ideal but good enough. In their absence, it would be much more plausible to say that space technologies really are untimely. And this applies even in the light of the many and well-known limitations and paradoxes of liberal democracy and of democracy of any sort.

The Earth already contains some very dangerous things and is likely to contain an increasing number of them over time. But what this points towards is the need to be less of an enclosed system. By which I mean that there is a growing need for us to be able to remove things from the system. Figuratively, we might think of this as allowing the planet to breathe: to take in the things required for life but also to remove the dangerous or threatening by-products. Some obvious examples would be nuclear waste, sequestered CO_2, and biohazardous materials that we may need to experiment on, but which will leave us globally vulnerable if we carry out such experimentation on the planet's surface or store the materials here rather than elsewhere. While some of these materials could safely be pushed into space using some approximation to current technologies, we cannot use anything like our current rocket technologies to remove nuclear waste in a safe manner. We would, instead, need something functionally equivalent to a space elevator to do so. By which I do not mean that we actually need a space elevator (predicting future technologies in any detail is fraught with difficulties) but rather that there is a growing need for something safe and *functionally equivalent*. Now looks like a very good time to start developing such a capability.

Perhaps there will be a better moment to do so further down the line. But we cannot know this, and it may simply not be the case. What we need is a moment that is *good enough* even if non-ideal. And if this is the case then space technologies are *timely enough* even if not *timely* simpliciter. Of course, if one believes that we can simply stop generating toxic materials then there will be no need to allow the Earth to breathe through the continuing development of space

174 Tony Milligan

technologies which might eventually allow for safe removal. But that is a picture not only of a renunciation of space ambitions, but the renunciation of a good deal of terrestrial technology such as biotechnology and the nuclear technology on which we are becoming increasingly dependent. And for a multiplicity of reasons, some of them political, this seems like an unlikely outcome. Apart from anything else, nobody is in a position to bring it about. Nobody is globally in charge in a way that might make it possible.

Acknowledgements

This article is part of a project that has received funding from the European Research Council (ERC) under the European Union's Horizon 2020 research and innovation programme (Grant agreement No. 856543).

References

Arendt, H. (1963). The conquest of space and the stature of man. In H. Arendt (Ed.), *Between past and future*. Penguin.

Ballard, J. G. (1981). Memories of the space age. In J. G. Ballard (Ed.), *The complete short stories*. Harper Perennial.

China Space Report. (2024). Wenchang space launch centre. https://chinaspacereport. wordpress.com/about-us/

Coleman, N. (Ed.) (2022), *Military space ethics*. Howgate.

De Groot, G. (2006). *The dark side of the moon: magnificent madness of the American lunar quest*. Vintage.

Deudney, D. (2020). *Dark skies: space expansionism, planetary geopolitics, and the ends of humanity*. Oxford University Press.

Dunleavy, H. (2021). An indigenous group's objection to geoengineering spurs a debate about social justice in climate science. Inside Climate News. https:// insideclimatenews.org/news/07072021/sami-sweden-objection-geoengineering-justice-climate-science/

Etzioni, A. (1964). *The Moon-doggle: domestic and international implications of the space race*. Doubleday.

Garan, R. (2015). *The orbital perspective: An Astronaut's View*. Metro.

IWGIA. (2023). Sámi activists demand removal of wind turbines in Fosen. https:// www.iwgia.org/en/news/5278-press-fosen-oct2023.html

Latour, B. (2017). *Down to Earth: politics in the new climatic regime*. Polity.

Leishner, R. B., & Hogan, T. (2019). *The view from space: NASA's evolving struggle to understand our home planet*. Kansas University Press.

Lovelock, J. E. (1965). A physical basis for life detection experiments. *Nature*, *207*(4997), 568–570. https://www.nature.com/articles/207568a0.pdf

Malcolm, X. (1963). The house negro and the field negro. https://www.youtube.com/ watch?v=7kf7fujM4ag

Milligan, T. (2015). *Nobody owns the Moon: The ethics of space exploitation*. Jefferson NC: McFarland and Company.

Milligan, T. (2023a). Inclusion and environmental protection in space. *The Brown Journal of World Affairs*, *29*(1), 116–128.

Milligan, T. (2023b). Is space expansion the road to dystopia? *Ethics and International Affairs*, *37*(4), 470–489.

Mumford, L. (1957). *The Transformations of Man*. Allen & Unwin.

NASA. (2010). The deadly Van Allen Belts. https://www.nasa.gov/wp-content/uploads/2010/06/smiii_problem7.pdf

Redfield, P. (1996). Beneath a modern sky: Space technology and Its place on the ground. *Science, Technology & Human Values*, *21*(3), 251–274. https://doi.org/10.1177/016224399602100301

Redfield, P. (2000). *Space in the Tropics: from convicts to rockets in French Guiana*. University of California Press.

Smith, K. N. (2018). The correction heard 'round the world: When the New York Times apologized to Robert Goddard. *Forbes Magazine*. https://www.forbes.com/sites/kionasmith/2018/07/19/the-correction-heard-round-the-world-when-the-new-york-times-apologized-to-robert-goddard/

Toynbee, A. J. (1976). *Mankind and Mother Earth*. Oxford University Press.

White, F. (2023). *The overview effect: Space exploration and human evolution*. Multiverse.

Wibeck, V., Hansson, A., & Anshelm, J. (2015). Questioning the technological fix to climate change: Lay sense-making of geoengineering in Sweden. *Energy Research & Social Science*, *7*, 23–30. https://doi.org/10.1016/j.erss.2015.03.001

Wildman, L. (2013). Dam removal: A history of decision points. In J. De Graff & J. E. Evans (Eds.), *The challenges of dam removal and river Restoration* (pp. 1–10). Geological Society of America.

Chapter 9

Moral Vistas to Xenic Beyonds: Fostering Moral Imagination to Pre-empt Monsterization in Future Encounters With Extraterrestrial Life

George Profitiliotis

Blue Marble Space Institute of Science, USA

Abstract

Outer space has been described as a liminal landscape. As such, it appears to have an intricate connection to the concept of the monster, which is a powerful metaphor that reflects and demarcates the culture that constructs it. In this vein, the scientific search for extraterrestrial life has been previously argued to possess characteristics of monstrosity. Consequently, the object of this search, that is, alien life, also bears the mark of the monster, as it can effortlessly tap into the particular leitmotifs of the monstrous that are largely constant across cultures, despite the situatedness of the monster. This fore-shadows a risk of monsterization of the parties involved in a future discovery of extraterrestrial life, that is, the monsterization of both the humans and the extraterrestrial life. In turn, this can adversely affect moral appraisals in future encounters with extraterrestrial life by rendering theoretical ethical approaches ineffective, as monsters are not only always outside the moral order but essentially defy and transgress it. Grounded on the intertwinement of moral imagination and anticipation and drawing on the inherent educational power of monsters, this chapter offers a theoretical exploration and a practical intervention in the form of a novel futures literacy workshop to help pre-emptively decrease the potential for the monsterization of humans and extraterrestrial life in the case of a future discovery. This contribution is in line with pragmatist ethics and is envisioned as a preparatory, complementary pedagogical approach to the traditional teaching of ethical theories and applied ethics to university students.

The Ethics Gap in the Engineering of the Future, 177–199

Copyright © 2025 George Profitiliotis

Published under exclusive licence by Emerald Publishing Limited

doi:10.1108/978-1-83797-635-520241010

178 *George Profitiliotis*

Keywords: Monster; extraterrestrial life; anticipation; moral imagination; futures literacy; liminality

Introduction: Outer Space, Monsters, Aliens, Anticipation and Moral Imagination

Outer space is not simply an area; rather it is an 'always liminal' landscape that is 'both a conceptual, imaginative realm and a literal physical place' (Walkowicz, 2023). Being liminal, outer space occupies a blurry region beyond all existing social order, albeit with rather permeable boundaries. Always in flux, this borderland is constructed by terrestrial societies that imagine it, and in turn constructs them dialectically by offering a means of demarcation (Dickens & Ormrod, 2016). Indeed, since antiquity and up to contemporary times, a culture's understanding of the cosmos and its relation to it has influenced social and political structures, as well as elements of a culture's worldview, including those towards other cultures but also those that define what it is to be human to that culture (Treviño, 2023). This bidirectional relationship between outer space and terrestrial societies carries fears and hopes related to humans accessing space and oftentimes to 'alien Others' accessing Earth (Dickens & Ormrod, 2016). In fact, the sheer enormity of outer space has been shown to induce effects that lie at both the positive and the negative ends of the ambiguously valenced experience of awe (Hornsey et al., 2018). Arguably, then, outer space as a liminal landscape appears to bear strong connections to the concept of the monster, that is, liminality incarnate. The 'monster', as operationalized in 'monster theory', is a liminal entity that reflects and demarcates the culture that constructs it (Cohen, 1996); monsters brought forward by science and technology, in particular, embody novel and ambiguous phenomena which fit simultaneously two preconditioned cultural categories that were originally thought to be mutually exclusive, thereby challenging the right symbolic order and evoking ambivalent feelings of fear and fascination (Smits, 2006). Consequently, the monstrosity of the search for extraterrestrial life, as discussed elsewhere (Profitiliotis & Theologou, 2023), is not an isolated case but rather a substantially potent manifestation of the deeper undercurrent of outer space's liminality. Further, seeing a potential discovery of extraterrestrial life – including intelligence – through the lens of the monster is a straightforward extension of the aforementioned: it can be conceptualized as the outcome of a monstrous endeavour; a threat to the existing symbolic order; a novel and ontologically liminal event whose encounter can catalyse the definition of new categories and concepts (Profitiliotis & Theologou, 2023).

Alien life is frequently framed as a monster in cultural products, especially in popular culture. In particular, this phenomenon of monsterization has already been noticed in the anticipations of intelligent extraterrestrial life which appear to be ingrained in the grooves of 'angelic' and 'demonic' extraterrestrials (Green, 2022), or the 'celestial saviour model' and the 'alien enemy model' (Peters, 2010). Importantly, though, analysing such anticipations of extraterrestrial life through the prism of monster studies allows the utilization of insights gained by empirically studying various historical instances of monstrosity across cultures, serving as a proxy to the unprecedented event of a successful discovery. Specifically, despite their cultural specificity that is intertwined with their definition, there are three interrelated leitmotifs related to monsters that are largely constant across cultures (Asma, 2009): the monstrous geography, the monstrous appearance, and the monstrous encounter. In terms of geography, the two interrelated principles that traditionally determined the places 'normally' and frequently inhabited by monsters were the extremes of temperature and the remoteness from the known world, as demonstrated, for example, by the mediaeval fascination with the 'monstrous races'. The element of remoteness was relative and contingent to what was considered 'known' or familiar, ranging from a village to a continent. Interestingly, the density of monsters in the familiar world was assumed to be lower because of their clash with 'civilization', e.g. heroes that exterminated them, which rendered the remainders exotic marvels to be overpowered and brought to the known world as tamed curiosities (Van Duzer, 2016). Both extreme environmental conditions and physical distance are fundamental aspects of the search for alien life; for some, the idea of supposed, elusive alien visitations also corresponds to rare curiosities amidst the familiar world. In terms of appearance, physical deformity, abnormality, and ugliness present across a population, rather than just a few individuals, were the principles that determined a monster's morphology, as demonstrated, for example, by the Cyclopes (Davies, 2016). An illustrative example that demonstrates the intrinsic link between physical appearance and monstrosity is the derivation of the Ancient Greek adjectives 'πελώριος' and 'τεράστιος', both meaning enormous, from the nouns 'τέρας' and 'πέλωρ' that mean 'monster'. Indeed, recent empirical research has revealed that exceptionally large animals trigger the simultaneous activation of fear and admiration emotions in humans (Prokop et al., 2023), which is a fundamental attribute of monsters (Clasen, 2012). As part of an alien organism's phenotype, appearance will be fundamentally influenced by evolutionary processes taking place in peculiar extraterrestrial environments, potentially resulting in aesthetic features that could correspondingly bias human emotions. Lastly, in terms of encounter, monsters historically generated the effects of wonder and terror, usually 'encountered first as abominations, then with sympathy, then as projects to fix'; however, it is possible for an encounter to break free from the tendency to cast out the monster as an abnormality, and to become productive by offering the space to redefine human potentiality (MacCormack, 2020). When the monster is not simply expelled as an anomaly, such a 'transformation and/or loss of self' of the human is induced by a forced reconsideration of familiar worldviews that challenges anthropocentric conceptions of the universe and by an unnerving recasting of oneself through the gaze of the Other. The effect of this encounter can be accentuated by the biological remoteness, the evolutionary alienation, of the monstrous Other from the human (Gunderson, 2017). Looking for anomalies and signs of abnormality with respect to the context is a key pillar of the search for alien life. Furthermore, the biological remoteness of a potential alien organism from the human can be reasonably presumed to be vast, influencing the effect of the encounter accordingly. The aforementioned arguments, alongside the observation that the monstrous

180   George Profitiliotis

geography, the monstrous appearance, and the monstrous encounter have all been historically used as moral ploys to justify the annihilation of monsterized Others (Asma, 2009), paint a bleak future for the moral appraisal of an encountered extraterrestrial organism.

As more and more actors become involved in space exploration and utilization, particularly non-state ones, the potential for discovering extraterrestrial life and the diversity of particular discovery situations increase significantly, given the fact that discoveries can occur both intentionally and unintentionally. Hence, for the sake of resilience, preparing the engineers, scientists, and space professionals of tomorrow for a future hypothetical involvement of theirs in an ethically problematic situation of encountering extraterrestrial life cannot rely solely on speculative case studies that are limited by underlying assumptions bounded by the extent of what is currently imaginable. In other words, a well-rounded preparation for such a future hypothetical situation cannot afford to overlook the crucial first step in moral reasoning that precedes moral judgement, that is, moral sensitivity. Moral sensitivity is the interpretation of a situation as morally relevant (Rest, 1994), the recognition of a moral issue; its three features are identifying courses of action, determining who warrants moral concern, and perspective taking, all of which can be enhanced by moral imagination (Moberg & Seabright, 2000). The process of determining who warrants moral concern, in particular, has been proven historically to be vulnerable to distortion in the case of monsterized Others, such as outsiders, strangers, and outgroups, resulting in their exclusion from the scope of moral concern (Monteiro et al., 2023). This means that a future encounter with extraterrestrial life carries a looming danger of descending into moral exclusion, obstructing sound moral judgement due to its capacity for triggering the effect of monstrosity; to phrase it differently, the ethical dimension of such a discovery could be completely overlooked. The goal of this chapter is to present a novel participatory workshop aimed at sharpening the moral sensitivity of university students in the space sciences, engineering, and other disciplines, by means of augmenting their moral imagination, to prevent the monsterization of humans and extraterrestrial life in the case of a future discovery, which could otherwise render theoretical ethical approaches ineffective. This chapter acknowledges the idea that an encounter with extraterrestrial life can stretch and test the standard theoretical foundations of ethics (Green, 2014) and offers a preparatory, complementary pedagogical intervention to help pollinate the students' capacity for anticipation and moral imagination.

Anticipation and moral imagination have been described as two key interrelated elements of engineering ethics education (van Grunsven et al., 2023). In general, anticipation can be defined as a forward-looking attitude that informs action in the present (Poli, 2017). Definitions for moral imagination vary (van Grunsven et al., 2023), although its core essence is succinctly captured in Johnson's formulation: the 'ability to imaginatively discern various possibilities for acting within a given situation and to envision the potential help and harm that are likely to result from a given action' (Johnson, 1993). The 'attentiveness to more than is immediately visible', the 'creative act', and the 'quality of transcendence' captured in Johnson's formulation were carried on into Lederach's more relationally- and less analytically-oriented formulation, which further illuminates the intertwinement of moral imagination with anticipation: 'the capacity to imagine something rooted in the challenges of the real world yet capable of giving birth to that which does not yet exist' (Lederach, 2005). In this formulation, moral imagination consists of four required elements: the ability to imagine oneself 'in a web of relationship even with their enemies'; the entertainment of 'a paradoxical curiosity' that surpasses 'forced containers of dualism and either-or categories'; the provision of space for 'the appearance of the creative act' that pushes 'the edges of what is thought to be real and possible'; and the willingness to 'take a risk' and step 'into the unknown' and into 'mystery' (Lederach, 2005). After being erroneously neglected and marginalized for most part of the history of moral philosophy and moral psychology, moral imagination is now seen not as simply subsidiary to the rational application of ethical theories, but as constitutive of moral understanding, as it facilitates the assessment of a moral problem, the projection of possible solutions, and the simulation of a situation's outcomes according to certain values (Johnson, 2016). Thus, enhancing moral imagination can be seen as a way of broadening democratic deliberation in anticipation of future science and technology issues, in line with a pragmatist ethics approach that sees values not as universal and transcendent but as dynamic, collective, workable solutions to practical problematic situations in particular cultures, practices, times, and spaces (Lucivero, 2016).

As demonstrated in a rich diversity of bioethics cases, the employment of the monster metaphor appears capable of influencing – constructively or destructively – moral imagination and discourse, as it possesses moral connotations that can affect the process and scope of ethical deliberation in the everchanging context of symbolic boundaries that flex and shift due to scientific and technological innovations (ten Have, 2022). Given how a 'failure of imagination' in astrobiology (Dick, 2018) could adversely affect moral appraisals in future encounters with extraterrestrial life, this chapter makes the theoretical case and offers a practical intervention for fostering moral imagination via an ex ante constructive exploration of crafted monsters to pre-empt the destructive monsterization of such situations.

The Monster: an Inhabitant of Moral Imagination and a Catalyst of Transition

Monsters have a deep and essential connection to moral imagination. Unpacking and understanding this connection can reveal ways to operationalize the monstrous in practical interventions that aim to enhance moral imagination. The character of signs, demonstrations, portents and omens that monsters possessed in antiquity was an essential justification of their imagined existence: the purpose of monsters was to teach moral lessons, by means of either exhibiting past or present transgressions or warning against future ones (Gilmore, 2003b). In Cohen's approach, monsters retain their fundamental pertinence to morality, although not from a supernatural perspective: the fifth thesis of monster theory reflects this aspect by suggesting, in paraphrase, that the monster demarcates the limits of knowing within a culture and delimits the borders that must not be crossed. These limits and borders refer not only to physical geography but also to conceptual and cultural territory. In that sense, the monster acts as a force of prohibition by announcing what menaces lurk beyond the permissible, what fiendish metamorphoses can afflict trespassers, or sometimes both. This active patrolling of the boundaries of culture means that acting against prescriptions may only occur through the body of the monster (Cohen, 1996). Historically, this recipe has also been used to monsterize various kinds of deviants and Others, humans and non-human animals alike: sometimes they were labelled as hideous non-conformers who deserved to fall prey to even worse monsters in cautionary tales, other times as kindred to bona fide monsters, and at times even as true monstrous beings in themselves. Indeed, the monster's embodiment of this unique and exclusive permission to transgress the established symbolic order has been so potent that it has been leveraged as both a devaluing ploy to justify discrimination and violence by dominant culture groups against outsiders, and as a unifying banner embraced willingly by the marginalized to serve as an emblem of their resistance and their emancipation from the iron hand of their oppressors. Arguably, then, the monster may be conceptualized as a moral heuristic, a short-cut in information processing to offer rule-of-thumb guidance in terms of what should be done or avoided: monstrous deeds are incongruent with being human.

The evocation of the monster as a moral heuristic can be further argued to function in both an autonomous way, i.e. intuitively, and in a consciously controlled way, i.e. deliberatively, in line with dual-process approaches to information processing for moral decision-making. Specifically, such dual-process approaches conceptualize information processing as comprising two qualitatively distinct types of processes supported by a set of multiple cognitive or neural systems: Type 1 processing is autonomous, working without conscious control, and is often described as fast, effortless, associative, and parallel; Type 2 processing is primarily consciously controlled, albeit operating in a continuum from weakly controlled to strongly controlled, and is often described as slow, effortful, deductive, and serial. Depending on the kind of heuristic, the person, and the context, moral heuristics are not limited to only one type of processing, since simplification can be intuitive but can also be reflective: some can operate at the level of one or the other type, others at the level of one type but not the other, and others at the level of one and the other type simultaneously (Nadurak, 2022). As discussed in the previous paragraph, the deliberate monsterization of humans, non-human animals, places, things, and other types of entities in a broad sense can be seen a tactic to deprive a target from its moral standing (Moser, 2020), while the deliberate self-monsterization of the tyrannized can be seen as an insurgency against their tyrants (Nordmarken, 2014). In that sense, this deliberate evocation of the monster as a moral heuristic appears to be a formidable tool for both maintaining and challenging power structures. However, the conscious use of this tool still draws its might from the intuitive influence of the monster while further amplifying it. To phrase it differently, monsters appear to function as a moral heuristic operating simultaneously at the level of Type 1 and Type 2 processing.

As such they can play a key moral role in shared imaginaries, including but not limited to social, political, and cultural ones, and this role has a mutually reinforcing relationship with the monsters' dependent and necessary position in human imagination as the essential antitheses to the humanness of their imaginers (Jacob & Bernardi, 2019).

The fundamental connection between the human and the monster as a moral heuristic is a constructive one (Coeckelbergh, 2020): in the realm of moral imagination, the monster acts as a virtual sparring partner, a beneficial foe, a symbol of instability and crisis that can help humans illuminate the manifold facets of hypothetical challenges and rehearse potential responses to them. Operating as persistent imaginary contenders, monsters stand in for potential future threats that may require active resolution, and for potential future trials that may test unquestioned core values and reveal moral truths. In this respect, the moral heuristic of the monster supports the proactive sensemaking of troubling situations before they occur, thereby assisting humans in their efforts to imagine and realize the 'good life' by means of both preparatory reasoning and preparatory feeling (Asma, 2020). In fact, such is the gravity it exerts to the appreciation of human life – both in terms of understanding it and in terms of valuing it – that, in certain special times, the monster is not only mentally accessed by means of thinking and feeling but is also embodied and performed. Times of festival, e.g. Carnival or Halloween, are liminal times, times out of time that are not everyday life nor completely outside of it, which renders them particularly suitable for incorporating the monster, even in this contemporary era. Festivals constitute a suspension of the daily routine and a temporally bounded pocket during which the playful transforms the mundane, giving rise to an ambiguous alternative reality wherein possibilities are expanded, social categories are rearranged, and creative social experimentation is celebrated (Gabbert, 2022). The examples of Carnival and Halloween vividly demonstrate how festivals ritualistically overturn normative cultural conventions to allow otherwise 'normal' people to experience joyful deviance by temporarily becoming monsters (Benshoff, 2020). In the safety of the festival's liminality, the monster is externalized, becoming a conduit to escape constraints and to express fantasies of performing or engaging with the morally forbidden, as long as it does not become a threat (Cohen, 1996). Interesting implications of this fluidity of moral boundaries (Michael, 2021) can be drawn from the fact that the boundaries that separate the festival's in-betweenness from the daily routine have been shown to be quite permeable, affording the opportunity for leaks of elements of transformation into the ordinary reality (Gabbert, 2022).

This performance of the monstrous during the liminality of festive periods highlights one more connecting thread between the monster and moral imagination: rites of passage. Rites of passage are a particular kind of ritual incorporating liminality in its most strict sense: they mark moments of a person's transition from one social category to another via a threshold of in-betweenness. It is this bona fide liminality that appears to make rites of passage even more fitting to host the monster – that is, liminality incarnate – than the more broadly defined liminal times of festivals. In their fundamentally ambiguous nature that

184   George Profitiliotis

bridges and transgresses mutually exclusive categories – such as being capable of both good and evil, being both primitive in form and just or wise in counsel, being both lethal and auspicious – , ritual monsters have the proper credentials to facilitate the transition of a person across boundaries. As a creature of in-betweenness, sometimes the monster acts as a warning, startling a person by showing them how an unaccomplished transition could render them social misfits and marginals. Other times, this same hybridity acts as a positive example of totality, exposing a person to a fused incarnation of supposedly mutually exclusive categories, e.g. both nature and culture, both animal and human, in order to familiarize their minds with a wide spectrum of experiences and reflections on aspects of reality taken for granted, before they make a solid transition into their new identity (Lada-Richards, 1998). With their first documented appearance in cave art dating back to the Paleolithic, ambiguous monstrous figures may have represented the mystery of life since prehistory. From the rituals of contemporary preliterate cultures to the still persisting festivities of rural areas in Europe, embedded monstrous images function didactically, encouraging the subjects of the ritual, especially young people, to clearly distinguish, reflect on, and understand the different elements of reality and the 'normal', as conceived in their culture. By embodying the unknown, the unimaginable, and the 'outside' of a particular culture – all aspects of the world that must be managed – , ritual monsters can vividly teach the moral traditions and awaken humans to the values of their culture; by dismantling the normal reality into its constituents and then recomposing them together into novel and compelling combinations, ritual monsters sometimes liberate and awaken the mind to conscious deliberation (Gilmore, 2003a).

One well-documented illustrative example of the ancient association of some monsters with rituals is the case of centaurs and gorgons in the iconography of Bronze Age Greece. Proto-centaur hybrids were already established as intermediaries to the divine in Minoan Crete; this helped them become positive mediator figures in social rites of passage in the Late Bronze Age, nurturing young boys to maturity – like the example of the wise centaur Cheiron – , emphasizing particular cultural values and teaching them cultural norms and social mechanisms. Gorgons, on the other hand, are thought to have functioned as a potent threat signalling the imminence of death in rites of passage, thereby leading young boys to a ritual rebirth in a new identity; it was their adversarial shock value that set the values and boundaries of culture and publicly reified social hierarchies and social order. Head-enveloping clay masks representing frightening gorgons have been argued to have been worn by the ceremonial antagonists of young boys in such rites. However, as transitional rites involved the temporally contained cancelation of values and hierarchies, masks were not only worn by others to shock the subjects of the ritual but in some cases also by the subjects of the ritual themselves, granting them the experience of separating from their past identity, social category, and status, and transitioning to the new one (Langdon, 2008). When worn by the subjects of the ritual, monstrous masks are thought to have enabled them to explore the marginality, strangeness, and otherness afforded by monstrosity, allowing them to subvert and break their culture's norms, rules, and categories so that they could learn to stabilize and internalize them in their lives (Lada-Richards, 1998).

Moral Vistas to Xenic Beyonds *185*

Consequently, it can be argued that some monsters – but not all of them – appear to reside as potent metaphors in human imagination, functioning as both intuitive and deliberate moral short-cuts. Usually, monsters can only be mentally accessed by means of thinking and feeling. However, being liminality incarnate, monsters can also be embodied and performed in certain contexts characterized by a broadly defined liminality, such as liminoid festivals, or by a true liminality, mainly rites of passage. At the same time, this liminality can become fertile ground for moral imagination, providing 'a period of scrutinization of the central values and axioms of the culture in which it occurs' (Turner, 1969) and 'a realm of pure possibility whence novel configurations of ideas and relations may arise' (Turner, 1967). Hence, the monster can be argued to reinforce all the four required elements of moral imagination in Lederach's formulation (Lederach, 2005): the ability to imagine oneself 'in a web of relationship even with their enemies' is served by the monster's dialectical role as a necessary antithesis and an opposition to the human that defines the boundaries of society and culture; the entertainment of 'a paradoxical curiosity' that surpasses 'forced containers of dualism and either-or categories' is served by the monster's hybridity and transgression of mutually exclusive categories that demands new ways of perceiving the world; the provision of space for 'the appearance of the creative act' that pushes 'the edges of what is thought to be real and possible' is served by the monster's embodiment of the unique and exclusive permission to question and overcome the established symbolic order that empowers transformation, especially in the context of liminal situations; and, lastly, the willingness to 'take a risk' and step 'into the unknown' and into 'mystery' is served by the monster's patrolling of the limits of knowing and its warning of the menaces that lurk beyond, which generates both fear and fascination.

Assisting a Moral Transition Into a World With Aliens: Unfamiliar Futures and Beneficial Monsters in Simulated Liminality

If liminality per se, as encountered mainly in rites of passage, and the monster, as liminality incarnate, can reinforce moral imagination, then interventions targeted at the scholars and professionals of tomorrow for facilitating their anticipation of encounters with extraterrestrial life might benefit from incorporating these elements into their plan of action. Several creative and arts-based approaches to cultivating moral imagination have been discussed in the literature, many of which have been specifically crafted for educating the scientists, engineers, and professionals of tomorrow. Examples include students reading short stories and novels (Carlson & Welker, 2001), particularly Mary Shelley's Frankenstein as a parable (Laugelli, 2023), students watching fictional films and series (Hitt & Lennerfors, 2022), students producing, performing, and watching drama (Birch & Lennerfors, 2020), students performing role-playing (Lennerfors et al., 2020),

186 George Profitiliotis

students engaging in futuristic simulations (Propst & Robinson, 2021), students building fictional design prototypes (York & Conley, 2020), and students immersing themselves in augmented reality-based behavioural simulations (Sari et al., 2021). In parallel, other published examples have shown that operationalizing the monster can fruitfully augment activities that aim at enhancing engagement with the future(s) (Hovorka & Peter, 2018), activities that aim at promoting contemplation of unforeseen and problematic future consequences of a technology (Dove & Fayard, 2020), and activities that aim at nurturing responsible innovation and teaching innovation ethics through future scenarios (Conley et al., 2023). A potentially powerful but underexplored type of moral imagination-fostering intervention that could benefit greatly from the integration of liminality and monsters are 'Futures Literacy Laboratory'-inspired workshops that can enhance anticipation and imagination.

As recommended elsewhere (Profitiliotis & Theologou, 2023), participatory 'Futures Literacy Laboratory'-inspired workshops could be used to bring together stakeholder groups that might potentially be affected by a future discovery of extraterrestrial life to help them anticipate more widely and diversely and build resilience in the face of unexpected future discoveries. In fact, such workshops appear well-suited to assist the participants' moral transition into a world with extraterrestrial life, by sharpening their moral sensitivity through augmenting their moral imagination, in anticipation of a potential future encounter. If embedded in a pedagogical context, such workshops may complement the teaching of ethical theories as part of the moral education of students in the space sciences, engineering, and allied disciplines, as well as in other fields. This section presents a novel futures literacy workshop design that aims at fostering moral imagination to help the scientists and engineers of tomorrow anticipate encounters with extraterrestrial life by pre-emptively decreasing the potential for the monsterization of humans and extraterrestrial life in the case of a future discovery. The overall workflow and the full content of this workshop will be presented hereinafter. It's worth noting that the workflow and content is designed to accommodate individual and collective work for 25 participants at maximum to keep the overall duration of the workshop limited at 2–3 hours. Adaptations are encouraged to accommodate particular needs. The title of this proposed workshop is 'Moral Vistas to Xenic Beyonds'.

The overall workflow of the 'Moral Vistas to Xenic Beyonds' workshop follows the three general phases of a Futures Literacy Laboratory-Novelty (FLL-N) (Miller, 2018). Firstly, the 'Reveal' phase helps participants surface the ideas about the future they hold in the present and interrogate the fundamental assumptions that are fuelling them. Then, the 'Reframe' phase exposes participants to an unfamiliar future that is completely liberated from these fundamental assumptions and helps them rehearse and make sense of a bewildering world. Finally, the 'Rethink' phase prompts participants to juxtapose their experiences in the previous two phases, to reflect on new aspects of the future that were discovered by this journey and to reassess the way they evaluate their actions in the present, which are implicitly framed by their future prospects. It should be noted here that the content of each phase has been designed for this particular

Moral Vistas to Xenic Beyonds **187**

application – although with minor modifications it may serve other similar applications too – . In addition, this workshop introduces a novel operationalization of Miller's six clusters of anticipatory assumptions (Miller, 2018) which are provided to the participants for use in the form of a 'List of Foundations', as demonstrated later in this section.

In particular, the content of the 'Reveal' phase of the 'Moral Vistas to Xenic Beyonds' workshop is structured as follows.

Initially, workshop participants are split into smaller groups of around five people. Within each smaller group, they are given the following instructions to work individually:

> Please close your eyes. It is now the year 2050. Imagine the search for alien life in its most likely state. In this most likely future of the search for alien life, imagine that you encounter in the news a widespread announcement that has something to do with the search for alien life. Please use words and simple sketches to recreate this announcement for the other participants. When recreating the announcement, try to illuminate its 'who', 'what', 'how', 'where', and 'why' aspects.

After the participants have created their announcements, they are given the following instructions to continue working individually:

> Reflect on the deeper attributes of the future world to which your announcement belongs. Which major assumption underlies it in terms of what you believe the future will be like? Write down your major assumption in the following format: 'I think this announcement is part of the likely future, because I believe that in 2050, compared to today, [something] will be [increased/decreased/the same/fundamentally different]'.

After the individual reflection, participants are given the following instructions to work as a smaller group:

> Take turns to present your announcement to your fellow participants and explain the major assumption that underlies it. While the presenter talks, listen closely and try to silently picture yourself experiencing this announcement right now and how it makes you feel. When a presentation of an announcement ends, vote on whether this announcement made you feel terror or amazement or awe (which is both terror and amazement) or none of the above.

After everyone has presented and voted on every announcement, participants are given the following instructions to continue working as a group:

188 George Profitiliotis

> Discuss, compare, and contrast the assumptions that underlie all your announcements freely and refrain from judging them. Overall, what do you think are the deeper foundations of your collective assumptions? Choose any that apply from the following 'List of Foundations' – you may select more than one: 'We built on forecasts, predictions, or projections that extrapolate from the past to the future'; 'We built on ideological or religious narratives that predetermine a fixed future destiny'; 'We speculated on future creative interventions to be undertaken by an individual or organization to solve some currently known issues in their external environment'; 'We speculated on future self-improvements to be undertaken by an individual or organization to overcome some currently known shortcomings within themselves'; 'We relied on a systematic consideration of the possible interplays of various factors to notice novel but recurring patterns as they were emerging'; 'We relied on active, alert awareness and mindfulness to notice something rare or unique as it was emerging'.

After the surfacing of the deeper foundations of their assumptions, the smaller groups are merged again in a plenary session and are given the following instructions to work as a larger group:

> Now that we have reconvened, each smaller group will take turns presenting only those announcements – if any – that have received a combination of votes on at least two of the three options, i.e., terror, amazement, awe (both terror and amazement), alongside the major assumptions that underlie them. We will gather these announcements and their underlying assumptions in the area designated as 'Monsters of the Probable Future by Present Standards'.

Afterwards, workshop participants are split again into their smaller groups. Within each smaller group, they are given the following instructions to work individually:

> Let's travel to the future once more. Please close your eyes. It is now the year 2050. Imagine the search for alien life but this time in its most desirable state. In this most desirable future of the search for alien life, imagine that you encounter in the news a widespread announcement that has something to do with the search for alien life. Please use words and simple sketches to recreate this announcement for the other participants. When recreating the announcement, try to illuminate its 'who', 'what', 'how', 'where', and 'why' aspects.

Moral Vistas to Xenic Beyonds   **189**

After the participants have created their announcements, they are given the following instructions to continue working individually:

> Reflect on the deeper attributes of the future world to which your announcement belongs. Which major assumption underlies it in terms of what you want the future to be like? Write down your major assumption in the following format: 'I think this announcement is part of my desirable future, because in 2050 I want [something] to be [increased/decreased/the same/fundamentally different], in comparison with today'.

After the individual reflection, participants are given the following instructions to work as a smaller group:

> Take turns to present your announcement to your fellow participants and explain the major assumption that underlies it. While the presenter talks, listen closely and try to silently picture yourself experiencing this announcement right now and how it makes you feel. When a presentation of an announcement ends, vote on whether this announcement made you feel terror or amazement or awe (which is both terror and amazement) or none of the above.

After everyone has presented and voted on every announcement, participants are given the following instructions to continue working as a group:

> Discuss, compare, and contrast the assumptions that underlie all your announcements freely and refrain from judging them. Overall, what do you think are the deeper foundations of your collective assumptions? Choose any that apply from the 'List of Foundations' – you may select more than one.

After the surfacing of the deeper foundations of their assumptions, the smaller groups are merged once more in a plenary session and are given the following instructions to work as a larger group:

> Now that we have reconvened, each smaller group will take turns presenting only those announcements – if any – that have received a combination of votes on at least two of the three options, i.e., terror, amazement, awe (both terror and amazement), alongside the major assumptions that underlie them. We will gather these announcements and their underlying assumptions in the area designated as 'Monsters of the Desirable Future by Present Standards'.

190   George Profitiliotis

Finally, the following instructions prepare the larger group of participants for the next phase of the workshop:

> Now, aware of the deeper foundations that sustain our engagements with our likely and desirable futures of the search for alien life and having catalogued some of their monsters, we leave them behind temporarily, as we continue our journey into the futures!

The content of the 'Reframe' phase of this workshop is structured as follows. Firstly, workshop participants are split again into their smaller groups. Within each smaller group, they are given the following instructions to work individually:

> Let's travel to the future for the third time. Please close your eyes. It is now the year 2050. However, the future is neither what we thought likely nor what we wanted. This future of the search for alien life is a strange one. The following is a key characteristic of this strange future world. The search for alien life is now seen as something both profoundly amazing and deeply terrifying, which causes people working in it to be seen as exotically intriguing and ominously menacing. As newcomers in this strange future of 2050, you encounter strange everyday things and strange everyday places. Here are some glimpses of places and things you recently encountered. All these places and things are somehow related to the search for alien life. Take a few moments to quietly reflect on these images.

For this activity, participants are provided with a set of less than 10 images in either physical or digital form, selected by the facilitator from the 'Moral Vistas to Xenic Beyonds' openly available living reference album: https://flic.kr/s/aHBqjB24xd.

After the individual reflection, participants are given the following instructions to work as a smaller group:

> Imagine you are in this future world and are encountering those places and things. Try to identify the emotion(s) you are experiencing while immersing yourself in this strange future world. Mark your emotion(s) on this collective Wheel of Emotions.

For this activity, participants are provided with Plutchik's 'wheel' of emotions (Plutchik, 2001) in either physical or digital form.

After everyone has marked their emotions, participants are given the following instructions to continue working as a smaller group:

> Keep in mind the images and your collective emotions. Imagine you are experiencing the people, the systems, the places, the things, and the events of this future world. What is going on there,

Moral Vistas to Xenic Beyonds **191**

particularly regarding the search for alien life? How does this world work? Write down your observations in this future world! Read what your fellow participants are observing and build on that!

For this activity, participants write down their observations in either physical or digital form in a way that allows everyone in their smaller group to read them.

After everyone has noted down their ideas about this strange future world, participants are given the following instructions to work individually:

As you are navigating a strange place in the midst of this future world, imagine that you encounter a widespread announcement that has something to do with a discovery of alien life. Encountering this announcement helps you explain why in this future world the search for alien life is seen as something both profoundly amazing and deeply terrifying, causing people working in it to be seen as exotically intriguing and ominously menacing. What is this announcement? How is it connected to the workings of this world and to the images and emotions you experience in it? What explanation does it offer for how the search for alien life is seen? Please use words and simple sketches to recreate this announcement for the other participants. When recreating the announcement, try to illuminate its 'who', 'what', 'how', 'where', and 'why' aspects.

After the participants have created their announcements, they are given the following instructions to continue working individually:

Reflect on the deeper attributes of the future world to which your announcement belongs. Which major assumption underlies it in terms of what you realized about this future? Write down your major assumption in the following format: 'I think this announcement is part of this strange future because, while experiencing this world, I realized that in 2050 [something] was [frequent/rare/non-existent]'.

After the individual reflection, participants are given the following instructions to work as a smaller group:

Take turns to present your announcement to your fellow participants and explain the major assumption that underlies it. While the presenter talks, listen closely and try to silently picture yourself experiencing this announcement right now and how it makes you feel. When a presentation of an announcement ends, vote on whether this announcement made you feel terror or amazement or awe (which is both terror and amazement) or none of the above.

192 George Profitiliotis

After everyone has presented and voted on every announcement, participants are given the following instructions to continue working as a group:

> Discuss, compare, and contrast the assumptions that underlie all your announcements freely and refrain from judging them. Overall, what do you think are the deeper foundations of your collective assumptions? Choose any that apply from the 'List of Foundations' – you may select more than one.

After the surfacing of the deeper foundations of their assumptions, the smaller groups are merged once more in a plenary session and are given the following instructions to work as a larger group:

> Now that we have reconvened, each smaller group will take turns presenting first those announcements – if any – that have received a combination of votes on at least two of the three options, i.e., terror, amazement, awe (both terror and amazement), alongside the major assumptions that underlie them. We will gather these announcements and their underlying assumptions in the area designated as 'Monsters of the Strange Future by both Present and Future Standards'. Each smaller group will then take again turns presenting the rest of their announcements and their major assumptions, and we will gather them in the area designated as 'Monsters of the Strange Future by Future Standards'.

Finally, the following instructions prepare the larger group of participants for the last phase of the workshop:

> Now, aware of the deeper foundations that sustain our engagement with a strange future of the search for alien life and having catalogued some of its monsters, we leave them behind temporarily, as we return to the present!

The content of the 'Rethink' phase of the proposed workshop is structured as follows.

Firstly, workshop participants are given the following instructions to continue working as a larger group:

> We are all back to today. Reflect on the previous activities by comparing and contrasting the deeper foundations that sustain the likely, the desirable, and the strange futures you collectively encountered. What pattern do you notice? Let's discuss!

After the collective reflection and discussion, participants should have noticed that the first two items from the 'List of Foundations' were the most prevalent when they engaged with the probable future, while the third and the fourth were the most prevalent in the desirable future, and the fifth and the sixth in the strange future.

Moral Vistas to Xenic Beyonds **193**

Following that, participants are given the following instructions to continue working as a larger group:

> Now that you have seen that thinking about different kinds of futures can be consciously harnessed by interrogating the deeper foundations that implicitly sustain them, let's turn to the 'Monsters of the Strange Future by Future Standards' catalogue we created. Observe the announcements in this catalogue and look for any patterns in light of the following questions. What is the Monster in those announcements? Which presumably distinct categories does the Monster blur? How is the Monster related to the Human? What are the established moral boundaries crossed by the Monster in that Strange Future? If these announcements were seen as both amazing and terrifying in that Strange Future, why didn't any of these announcements make you feel a combination of terror, amazement, and awe in the present? Why is the crossing of those moral boundaries not triggering the same effect in the present? Let's discuss these questions collectively!

After this reflection, participants are given the following instructions to continue working as a larger group:

> Let's turn now to the 'Monsters of the Strange Future by both Present and Future Standards' catalogue we created. Once more, observe the announcements in this catalogue and look for any patterns in light of the following questions. What is the Monster in those announcements? Which presumably distinct categories does the Monster blur? How is the Monster related to the Human? What are the established moral boundaries crossed by the Monster in both that Strange Future and the present? Why is the crossing of those moral boundaries triggering the same effect in the present? Let's discuss these questions collectively! What productive insights have we gained by comparing this kind of Monster with the previous one?

After the collective reflection and discussion, participants should have noticed that the cultural categories that order the world and define the 'proper symbolic order' are not rigid and can be reconfigured (Smits, 2006). The Monster by both present and future standards illuminates the cultural predisposition of rigid categories, whereas the Monster by future standards, which does not appear monstrous in the present, exposes participants to an easily accessible violation of this predisposition embodied in something that appears 'normal' in the present but is considered monstrous in the future. The former Monster can be seen as 'The Monster that Startles' – a gorgon – , while the latter as 'The Monster that Familiarizes' – a centaur – .

194 George Profitiliotis

Finally, participants are given the following instructions to continue working as a larger group:

> Let's turn now to the 'Monsters of the Probable Future by Present Standards' and the 'Monsters of the Desirable Future by Present Standards' catalogues we created. Once more, observe the announcements in these two catalogues and look for any patterns in light of the following questions. What is the Monster in those announcements? Which presumably distinct categories does the Monster blur? How is the Monster related to the Human? What are the established moral boundaries crossed by the Monster in the present? How can we harness the productive insights gained in the previous activity to scrutinize those cultural categories and moral boundaries of the present? How can we reconfigure some of those cultural categories and moral boundaries today to prevent the detrimental monsterization of humans and extraterrestrial life in the case of a future discovery? Let's discuss these questions collectively!

The goal of this final collective reflection and discussion is, first, to help participants operationalize the understanding that a culture's symbolic order is not rigid and, then, to apply it to a creative act of pushing the perimeter of consideration and considerability beyond its familiar territories, followed by a risk-defying, anticipatory reconfiguration of cultural categories and moral boundaries as a step into the unknown. Using the multiple kinds of monsters produced as moral vistas to both the future and the present is expected to help participants uncover the multifaceted alienation potential of encountering extraterrestrial life. In turn, this is expected to sharpen their moral sensitivity by broadening their moral imagination in anticipation of a future discovery that could otherwise challenge what unscrutinized cultural categories could afford to expect with detrimental implications.

Epilogue: Here Be Monsters

As an encounter with a prototypical Other, the discovery of extraterrestrial life – intelligent or not – may elicit a deep, gut-feeling reaction of monstrosity. The degree of this monstrous response could also be influenced by the aesthetic aspects of visual depictions accompanying this discovery across various communication channels, in both scientific images and artistic representations, by triggering relevant biases. The ensuing monsterization of the discovery by its beholders could, in turn, render theoretical ethical approaches ineffective, as monsters are not only always outside the moral order but essentially defy and transgress it. Moreover, the potential expansion of the veil of monsterization to individuals associated with the discovery or even with the search for extraterrestrial life more generally may endanger not only the status of this scientific endeavour but also the welfare and safety of humans, posing a catastrophic threat of an all-out 'moral panic' in the worst case. Needless to say, the scientists and engineers of tomorrow will not be immune to the allure of the monster. Concurrently, it is reasonable to assume that their role in future society will continue to make them key figures in the landscape of the macro-ethical dimensions of new science and technology, and the corresponding public dialogue. Therefore, preventing the aforementioned detrimental turn of events requires proactive interventions to complement the traditional teaching of ethical theories and applied ethics to university students in the space sciences, engineering and allied disciplines, as well as in other fields, by sharpening their moral sensitivity through fostering their moral imagination, in line with a pragmatist ethics approach.

To this end, this chapter introduced 'Moral Vistas to Xenic Beyonds', a participatory futures literacy workshop design focused on the theme of encountering extraterrestrial life. This workshop design was crafted specifically to enhance the participants' capacity for anticipation, by helping them cultivate their futures literacy, and to pollinate their moral imagination, by exposing them to the beneficial in-betweenness of the monster. Braiding the kindred threads of monsters, futures, and aliens, this intervention attempts to simulate the liminality of engaging with three different futures – the probable, the desirable, and the strange – , with the goal of easing the participants' moral transition into a future world where extraterrestrial life is a reality. Employing a learning journey of revelation, reframing, and rethinking across this host of different kinds of futures, this intervention strives to guide participants towards an understanding of the deepest implicit foundations that underpin their future imaginings, illustrating that they can be consciously harnessed to aspired ends. At the same time, drawing on the monster's heuristic potency, this intervention seeks to invigorate the participants' moral imagination by stimulating all four required elements in Lederach's formulation, that is, by helping participants untangle the relationship between the monster and the human in diverse arrangements of monsterization, by helping participants realize the artificiality of forced either-or categories via the directly accessible route of situations appearing normal today but deemed monstrous in the future, by providing participants with the necessary creative space for scrutinizing and pushing the edges of their culture's symbolic order, and by urging them to take the risk of reconfiguring present cultural categories and moral boundaries with a prudent eye towards the unknown and mysterious future.

In the vein of other pioneering pedagogical approaches that explicitly invite the monstrous into the classroom (Golub et al., 2017), this intervention aspires to channel the inherent educational power of monsters, albeit magnified through the lens of the futures, to help prepare the scholars and professionals of tomorrow for recognizing and pragmatically navigating the moral challenges of a potential future encounter with extraterrestrial life, the ultimate Other. If the desideratum of ethics education is to endow the decision-making toolkit of future citizens with a moral compass and a map of ethical problems, then the bequest of this intervention is a pair of binoculars for the mind: an instrument to push the perimeter of consideration and considerability beyond the familiar borders, into the 'hic sunt monstra' corners of the cultural territory, thereby affording the reshaping of the symbolic landscape itself.

References

Asma, S. T. (2009). Epilogue. In *On monsters: An unnatural history of our worst fears* (pp. 278–284). Oxford University Press.

Asma, S. T. (2020). Monsters and the moral imagination. In J. A. Weinstock (Ed.), *The monster theory reader* (pp. 289–294). University of Minnesota Press.

Benshoff, H. (2020). The monster and the homosexual. J. A. Weinstock (Ed.), *The monster theory reader* (pp. 226–240). University of Minnesota Press.

Birch, P., & Lennerfors, T. (2020, October). Teaching engineering ethics with drama. In *2020 IEEE Frontiers in Education Conference (FIE)* (pp. 21–24), Uppsala, Sweden.

Carlson, W. B., & Welker, R. (2001, June). The whammy line as a tool for fostering moral imagination. In *Albuquerque, New Mexico, American Society for Engineering Education 2001 Annual Conference* (pp. 24–27).

Clasen, M. (2012). Monsters evolve: A biocultural approach to horror stories. *Review of General Psychology, 16*(2), 222–229.

Coeckelbergh, M. (2020). Monster anthropologies and technology: Machines, cyborgs and other techno-anthropological tools. In D. Compagna & S. Steinhart (Eds.), *Monsters, Monstrosities, and the Monstrous in Culture and Society* (Vol. DE, pp. 353–369). Vernon Press.

Cohen, J. J. (1996). Monster Culture (Seven Theses). In J. J. Cohen (Ed.), *Monster theory: Reading culture* (pp. 3–25). University of Minnesota Press.

Conley, S. N., Tabas, B., & York, E. (2023). Futures labs: A space for pedagogies of responsible innovation. *Journal of Responsible Innovation, 10*(1), 2129179.

Davies, S. (2016). The unlucky, the bad and the ugly: Categories of monstrosity from the renaissance to the enlightenment. In A. S. Mittman & P. J. Dendle (Eds.), *The ashgate research companion to monsters and the monstrous* (pp. 49–76). Routledge.

Dick, S. J. (2018). *Astrobiology, discovery, and societal impact*. Cambridge University Press.

Dickens, P., & Ormrod, J. (2016). Introduction: The production of outer space. In P. Dickens & J. Ormrod (Eds.), *The Palgrave handbook of society, culture and outer space* (pp. 1–43). Palgrave Macmillan.

Dove, G., & Fayard, A.-L. (2020, April 25-30). *Monsters, metaphors, and machine learning*. CHI.

Gabbert, L. (2022). Monsters, Legends, and Festivals: Sharlie, winter carnival, and other isomorphic relationships. In D. J. Puglia (Ed.), *North American monsters: A contemporary legend casebook.* (pp. 298–314). University Press of Colorado.

Gilmore, D. D. (2003a). Ritual Monsters. In *Monsters: Evil beings, mythical beasts, and all manner of imaginary terrors* (pp. 155–173). University of Pennsylvania Press.

Gilmore, D. D. (2003b). Why study monsters?. In *Monsters: Evil beings, mythical beasts, and all manner of imaginary terrors* (pp. 1–10). University of Pennsylvania Press.

Golub, A., Hayton, H. R., & Jefferson, N. C. (2017). *Monsters in the classroom: Essays on teaching what scares us*. McFarland & Company.

Green, B. P. (2014). Ethical approaches to astrobiology and space exploration: Comparing Kant, Mill, and Aristotle. In J. Arnauld (Ed.), *Space exploration and*

ET: Who goes there? special issue of ethics: Contemporary issues (Vol. 2, pp. 29–44). *ATF Press.*

Green, B. P. (2022). The Search for Extraterrestrial Intelligence. In *Space Ethics* (pp. 153–170). Rowman & Littlefield.

Gunderson, M. (2017). Other ethics: Decentering the human in weird horror. *Kvinder, Køn & Forskning, 26*(2–3), 12–24.

Hitt, S. J., & Lennerfors, T. T. (2022). Fictional film in engineering ethics education: With Miyazaki's the wind rises as exemplar. *Science and Engineering Ethics, 28*(5), 44.

Hornsey, M. J., Faulkner, C., Crimston, D., & Moreton, S. (2018). A microscopic dot on a microscopic dot: Self-esteem buffers the negative effects of exposure to the enormity of the universe. *Journal of Experimental Social Psychology, 76*, 198–207.

Hovorka, D. S., & Peter, S. (2018). Thinking with monsters. In U. Schultze, M. Aanestad, & M. Mähring (Eds.), *Living with monsters? Social implications of algorithmic phenomena, hybrid agency, and the performativity of technology. IS&O 2018, IFIP advances in information and communication technology* (Vol. 543, pp. 159–176). Springer.

Jacob, F., & Bernardi, V. (2019). Introduction: All around monstrous or a critical insight into human-monster relations. In V. Bernardi & F. Jacob (Eds.), *All around monstrous: monster media in their historical contexts* (pp. v–xvi). Vernon Press.

Johnson, M. (1993). *Moral Imagination.* University of Chicago Press.

Johnson, M. (2016). Moral imagination. In A. Kind (Ed.), *The Routledge handbook of philosophy of imagination* (pp. 355–367). Routledge.

Lada-Richards, I. (1998). Monsters and monstrosity in Greek and Roman culture. In C. Atherton (Ed.), *Foul monster or good savior? Reflections on ritual monsters* (pp. 41–82). Levante Editori.

Langdon, S. (2008). Geometric art comes of age: An Archaeology of maturation. In *Art and identity in dark age Greece, 1100-700 BCE* (pp. 56–125). Cambridge University Press.

Laugelli, B. J. (2023, June). *Rogue Engineering: Teaching Frankenstein as a Parable of (Un)ethical Engineering Practice Paper.* Baltimore, MD. In *2023 ASEE Annual Conference & Exposition* (pp. 25–28).

Lederach, J. P. (2005). *The moral imagination: The art and soul of building peace.* Oxford University Press.

Lennerfors, T. T., et al. (2020, October). A Pragmatic Approach for Teaching Ethics to Engineers and Computer Scientists. In *2020 IEEE Frontiers in Education Conference (FIE)* (pp. 21–24). Uppsala, Sweden

Lucivero, F. (2016). Scenarios as "Grounded Explorations". Designing tools for discussing the desirability of emerging technologies. In *Ethical assessments of emerging technologies: Appraising the moral plausibility of technological visions* (pp. 155–190). Springer.

MacCormack, P. (2020). Posthuman teratology. In J. A. Weinstock (Ed.), *The monster theory reader* (pp. 522–539). University of Minnesota Press.

Michael, M. (2021). Aliens, monsters, and beasts in the cultural mapping of Nollywood cinematography. *Humanities Bulletin, 4*(1), 168–201.

Miller, R. (2018). Sensing and making-sense of Futures Literacy: Towards a Futures Literacy Framework (FLF). In R. Miller (Ed.), *Transforming the future: Anticipation in the 21st century* (pp. 15–50). Routledge.

Moberg, D. J., & Seabright, M. A. (2000). The development of moral imagination. *Business Ethics Quarterly, 10*(4), 845–884.

Monteiro, B., West, B., & Pizarro, D. A. (2023). Monsters and the moral psychology of the "other". In H. Kapoor & J. C. Kaufman (Eds.), *Creativity and morality* (pp. 161–173). Academic Press, an imprint of Elsevier.

Moser, K. (2020). J. M. G. Le Clézio's Defense of the Human and Other-than-human Victims of the Derridean "Monstrosity of the Unrecognizable" in the Mauritian Saga Alma. In K. Moser & K. Zelaya (Eds.), *The Metaphor of the Monster: Interdisciplinary approaches to understanding the monstrous other in literature* (pp. 63–84). Bloomsbury Academic.

Nadurak, V. (2022). Moral heuristics and two types of information processing. *Acta Baltica Historiae et Philosophiae Scientiarum, 10*(2), 46–62.

Nordmarken, S. (2014). Becoming ever more monstrous: Feeling transgender in-betweenness. *Qualitative Inquiry, 20*(1), 37–50.

Peters, T. (2010). ET: Alien enemy or celestial savior? *Theology and Science, 8*(3), 245–246.

Plutchik, R. (2001). The nature of emotions. *American Scientist, 89*(4), 344–350.

Poli, R. (2017). Introducing anticipation. In R. Poli (Ed.), *Handbook of anticipation: Theoretical and applied aspects of the use of future in decision making* (pp. 1–14). Springer.

Profitiliotis, G., & Theologou, K. (2023). The monstrosity of the search for extra-terrestrial life: Preparing for a future discovery. *Futures, 147*, 103117.

Prokop, P., Zvaríková, M., Zvarik, M., Provazník, Z., & Fedor, P. (2023). Charismatic species should be large: The role of admiration and fear. *People and Nature, 00*, 1–13.

Propst, L., & Robinson, C. C. (2021). Pandemic fiction meets political science: A simulation for teaching restorative justice. *PS: Political Science & Politics, 54*(2), 340–345.

Rest, J. R. (1994). Background: Theory and research. In J. R. Rest & D. Narvez (Eds.), *Moral development in the professions: Psychology and applied ethics* (pp. 1–26). Lawrence Erlbaum Associates.

Sari, R. C., Sholihin, M., Yuniarti, N., Purnama, I. A., Hermawan, H. D. (2021). Does behavior simulation based on augmented reality improve moral imagination? *Education and Information Technologies, 26*, 441–463.

Smits, M. (2006). Taming monsters: The cultural domestication of new technology. *Technology in Society, 28*, 489–504.

ten Have, H. (2022). Monsters. In *Bizarre bioethics: Ghosts, monsters, and pilgrims* (pp. 61–87). Johns Hopkins University Press.

Treviño, N. B. (2023). Coloniality and the Cosmos. In J. F. Salazar & A. Gorman (Eds.), *The Routledge handbook of social studies of outer space* (pp. 226–237). Routledge.

Turner, V. (1967). *The forest of symbols: Aspects of Ndembu ritual.* Cornell University Press.

Turner, V. (1969). *The ritual process: Structure and anti-structure.* Aldine Publishing Company.

Van Duzer, C. (2016). Hic sunt dracones: The geography and cartography of monsters. In A. S. Mittman & P. J. Dendle (Eds.), *The ashgate research companion to monsters and the monstrous* (pp. 511–565). Routledge.

van Grunsven, J., Stone, T., & Marin, L. (2023). Fostering responsible anticipation in engineering ethics education: How a multi-disciplinary enrichment of the responsible innovation framework can help. *European Journal of Engineering Education*, 1–16, Advance online publication.

Walkowicz, L. (2023). Foreword. In J. F. Salazar & A. Gorman (Eds.), *The Routledge handbook of social studies of outer space*. Routledge. xvii-xix.

York, E., & Conley, S. N. (2020). Creative anticipatory ethical reasoning with scenario analysis and design fiction. *Science and Engineering Ethics*, *26*, 2985–3016.

Chapter 10

Planning for the Future in Space – With or Without Radical Biomedical Human Enhancement?

Rakhat Abylkasymova[a] and Konrad Szocik[b]

[a]Independent Researcher, Poland
[b]University of Information Technology and Management in Rzeszow, Poland

Abstract

In our chapter, we want to point out the long-term ethical implications of the concept of space exploration and exploitation, which are usually overlooked today. Future space exploration and exploitation is assumed today as a certain part of human development and includes space tourism, scientific missions, space mining, as well as, in the further future, permanent settlement. But will not such long-term space exploration require the application of extraordinary solutions? In our chapter, we want to analyze this question with regard to the possible obligation or requirement to apply radical human enhancement. Among other things, we want to refer to the feminist perspective and also pay attention to issues such as exclusion and power structures. After all, it is impossible not to analyze the future of human beings in space without drawing attention to current capitalist exploitation of a global nature. We also point out that certain groups such as workers, women, and people with disabilities will be particularly vulnerable to exploitation and exclusion in space, and that human enhancement may negatively affect their social standing and empowerment.

Keywords: Space exploration; human enhancement; gene editing; capitalism; exploitation; feminism; planning

---

The Ethics Gap in the Engineering of the Future, 201–213
Copyright © 2025 Rakhat Abylkasymova and Konrad Szocik
Published under exclusive licence by Emerald Publishing Limited
doi:10.1108/978-1-83797-635-520241011

Introduction

Despite the fact that humans have been carrying out space missions for more than 60 years, the discussion on the ethics of space missions is often only about potential missions that may possibly come in the future (Schwartz and Milligan, 2016). Possibly, because the actual rationale in favor of space exploration is not so obvious and so strong to justify such a complex and expensive project. Paradoxically, then, we arrive at a situation where, when talking about the ethics of space missions, we are almost always talking about something that is not yet there and it is not known what it will be. This is not to say that those space missions that humans have been carrying out for decades are without an ethical component. However, not very much of the ethics of current space missions is different from the ethics of all other activities that humans carry out. Thus, when we talk about the ethics of space exploration, we have in mind such ethical problems that are specific only to this type of activity, possibly they will be more likely or more strongly visible in space.

Identifying potential invisible gaps in ethical thinking about our future in space, we turn our attention to the issue of biomedical human enhancement. Biomedical human enhancement includes procedures such as genetic modification, pharmacological agents, or computer–brain interfaces. The aim of these procedures is to improve the human being in order to make them more health resilient, fitter, stronger, smarter – in other words, to function better or as well as they did in the previous, standard environment (if the individual is moved to a more demanding environment). The idea of human enhancement expresses the desire to improve our lives and is a reaction to our concerns about the inconveniences of our bodies and our biology (Hauskeller, 2024). We also know that these procedures are controversial and usually either prohibited or strictly regulated. The environment of space missions is characterized by obvious difficulties and risks to human health and life, caused by cosmic radiation, altered gravity, living in a small isolated space, distance from Earth, constant dependence on the life support system. These are circumstances that create special justification for discussing the concept of radical biomedical human enhancement (Szocik & Braddock, 2019).

But even if we are willing to agree that a demanding environment can justify the use of exceptions to accepted rules and norms, we should also analyze the purpose and justification for the very space missions that would require radical human modifications. In our chapter, we want to discuss the bioethics of biomedical human enhancement for future space missions in the context of the justification for these missions. We thus turn our attention to a broader context that goes well beyond the moral evaluation of the procedure itself. We also consider which groups of future participants in space missions might be most vulnerable. Our paper points out that what is most relevant to the bioethics of space exploration in the context of ethical evaluation of human enhancement is the rationale for the mission and the goals set by the mission planners and organizers, that is, state agencies and private corporations. We reject those positions that analyze the controversiality of biomodification due to the structural intrinsic characteristics attributed to it, such as invasiveness or potential irreversibility. We reject counter-arguments that appeal to human dignity or the so-called human nature. Focusing on these metaphysical and structural arguments inherent in classical bioethics distracts from the core of the problem, namely the risk of exploitation and discrimination. Thus, in our chapter, we highlight only one counter-argument to the use of biomedical human enhancement, that is, the argument from capitalist and nationalist exploitation of humans and the cosmos. However, we do not ignore the arguments referring to safety, the assurance of which with regard to gene editing is highly problematic due to the unpredictability of the modifications introduced, as well as the risk of off-target effects (Gregg, 2023).

Human Enhancement and the Type of Space Mission and Its Rationale

An idea that recognizes the possibility of linking the justification for a controversial procedure to its purpose is moral exceptionalism if it is assumed that in certain circumstances, due to certain differences, different moral standards can be applied. Moral exceptionalism is relatively easy to apply to extreme environments such as space, military or Arctic expeditions. According to the logic of the ethics of extreme situations, under certain circumstances – extreme or unnatural precisely – other values than those generally accepted under standard conditions are considered most important (Garasic, 2021). The question that arises here is whether space exploration and exploitation justify the application of different ethical standards than the more or less acceptable moral consensus on Earth. Moral exceptionalism is in line with our intuitions. This is because we assume that different professional groups such as health care workers, services such as the police or soldiers may be exposed to more difficult, demanding situations and conditions than the rest of society. As part of our intuitions of moral exceptionalism, we also assume that in certain situations such as wars or epidemics and pandemics, it is also permissible to *exceptionally* just suspend certain rules and norms that we would not suspend were it not for these situations.

In the context of the rationale and goals of space exploration and types of space missions, it is worth starting with a disenchanted view of space exploration. The only credible type of space missions will be profit-oriented missions. This is evidenced by the dynamics of space exploration as we know it. We mean only human missions, because only such missions are concerned with the moral problem of using biomedical human enhancement. Only a profit motive can inspire state and/or private entities to pursue such projects. The only space mission of significance, that is, beyond Earth orbit, was the Apollo program missions carried out by the US, in which astronauts landed on the Moon. The only real purpose of these missions was political as part of the space race with the Soviet Union. But underneath the rather general and vague notion of political goals are specific, measurable economic and military objectives. In the case of the Apollo missions, economic goals did not mean profits from exploitation of the

Moon, but translating into potential economic gains on Earth and competing to become an Earth power dominant over others. The termination of the Apollo mission and the absence thereafter of any other human mission beyond Earth's orbit suggest that space exploration and exploitation, at least to the limited extent determined by technological capabilities, was not regarded as profitable either from the point of view of the state or the private sector. The situation changed after 2000, when the private space business emerged, opening the era of so-called New Space.

From today's perspective, we can conclude that the only real motivating factor for space exploitation and exploration is the benefits expected by private entrepreneurs. Thus, the only viable type of human space exploration will be profit-driven missions, carried out especially by private entrepreneurs, with some role and interest from the state (Rubenstein, 2022). In practice, human for-profit missions will involve sending workers to work in the space sector. These could be employees of corporations exploring the lunar surface, workers in the space mining sector focused on asteroid exploitation, or workers operating and at the same time living in cislunar, lunar, or Martian bases. The category of profit-oriented missions should include those organized by state agencies that combine economic interests with nationalistic and military ones.

The second viable type of mission, however possible in the distant future probably as a consequence of the aforementioned for-profit space exploitation phase, are missions aimed at settling humanity in space. This type of mission, however distant in time, is realistic and probable, but for reasons other than those traditionally discussed by the philosophy of space exploration and the discourse inherent in global catastrophes and risks studies framework. In the aforementioned approaches, the concept of space settlement is usually discussed in terms of saving the human species from an impending catastrophe on Earth. Space, therefore, is treated in terms of space refuge. However, the logic of the concept of saving humanity by settling in space is flawed for two reasons. First, it is not possible to save humanity in terms of numbers, only to allow a small fraction to relocate, which would save them from perishing by staying on Earth. For logistical and technical reasons, it is not possible to send a large number of Earth's inhabitants into space, much less all of them. This limitation casts doubt not only on the sensibility and feasibility of the concept but also removes the justification for future attempts to put it into practice. Secondly, the problem is the implementation of such a project over time. Deciding to build autonomous, self-sustaining shelters in space would require the ability to define and predict well in advance the catastrophe that is bound to happen and will certainly threaten humanity in such a way that survival on Earth would be impossible. It would therefore be necessary to be able to identify this type of catastrophe at least several decades before it occurs. It would also be a problem to identify the entity responsible for funding and preparing such a project. No less of a problem would be the selection of individuals who, if such a project were to be understood as a project of all humanity or carried out on behalf of all humanity, would have to be transparent. This would mean that more than eight billion people would know that the implementation of costly, time- and resource-consuming preparations for a settlement mission would only serve to save, for example, a maximum of a few dozen, possibly a few hundred thousand people. It is difficult to suppose that the public would approve the implementation of such a project. These and other counter-arguments serve to show that the idea of space settlement seen as a kind of salvation of humanity is unrealistic and certainly not a salvation of humanity.

However, this does not rule out the possibility of human settlement in space for other reasons. This process is likely to be a natural consequence of space exploitation. The space business will need workers in space, and they may have to live there. It is also possible that the development of space technology will lead to a situation where life in space will be only minimally ailing, and its attractiveness to at least some wealthy people will make life in space more desirable than life on Earth.

In this way, we were able to concretize and narrow the discussion of human future in space to the two most likely types of missions and activities: space business workers and permanent or semi-permanent space dwellers. The latter group will grow out of and at least partially overlap with the former. If we assume that the use of biomedical enhancement of participants in future long-term space missions will be medically feasible and recommended or even mandatory, then almost all of the people undergoing these modifications will be people in space for benefits. Significantly, these will be corporate employees living in harsh conditions and undergoing biomedical modifications for the benefit of corporate owners and politicians. Thus, we come to the point where we conclude that talking about biomedical human enhancement in space means employees of space corporations undergoing modifications for the profits of these corporations. An arguably small proportion may be people who choose space as a place of permanent settlement, assuming that life in space will for some reason be seen by these people as more attractive than staying on Earth. If we take a global perspective, and treat the future in terms of the increasing threats posed by climate change, the two types of space missions mentioned are morally questionable from the point of view of global problems and the most excluded and exploited people.

Given the classical approach exposing, among other things, the threat to so-called human nature, it is worth asking whether, if philosophers established that biomedical human enhancement does not threaten human nature, would it cease to be controversial even if it contributes to strengthening the dynamics of exploitation of at least the employees of space corporations? When adopting a classical perspective, the risk of exploitation and discrimination is usually treated as an additional social context that is not subject to proper bioethical analysis. In fact, only the perspective inherent in feminist bioethics makes the analysis and critique of mechanisms of discrimination and exploitation the central subject of its ethical analysis (Szocik, 2024).

Human Enhancement and Vulnerable People on Earth

Global and feminist bioethics draws attention to inequalities related to medical care, access to medical procedures, as well as the unequal effects of the application of particular medical procedures. The basic boundaries of inequality are set by race, gender, and socio-economic class. In countries with racist traditions, where racism is present at all levels, including structural and institutional, such as the USA, non-whites are treated worse, that is, they have more difficult access to medical care, than whites (Russell, 2022). It is also common in the West for middle-class whites to be treated as a default patient group by the health care system (Gregg, 2023).

A classic example that characterizes many countries is gender discrimination, where women are usually marginalized in health care compared to men. This includes a focus on research into drugs and procedures for men and the exclusion of women from clinical trials, a preference for men for expensive and technically advanced treatments, as well as gender biases regarding particular types of diseases. In the latter case, cardiovascular diseases are treated as the domain of men, while women's medical problems tend to be reduced to problems with their reproductive biology (Ballantyne, 2022). Discrimination and exclusion related to class characterize both societies with guaranteed universal access to health care and a business model specific to the USA. Where health care depends on privately paid expensive insurance, socio-economic class membership is critical. However, even with publicly-funded healthcare, its performance and quality is unpredictable, causing some patients, usually those with adequate financial resources, to seek help from private healthcare services.

In a structure as unequal as the health service, the possibility of biomedical human enhancement must reflect and replicate these inequalities. It can therefore be surmised that if various forms of biomedical modification by way of somatic and even germline gene editing were to be widely available, groups currently excluded from full access to all medical procedures may be similarly excluded from access to biomedical human enhancement. If gene editing is costly and privately funded by interested individuals, access to it would be analogous to the ability to access educational services where they are not publicly available. But class difference is not the only one that can affect accessibility to biomedical human enhancement. Another category potentially determining the availability of human enhancement could be gender. In an already highly gendered health care system, access to human enhancement may be determined by perceptions, but also expectations, associated with a particular gender. This may vary from society to society depending on their guaranteed acceptance of women's rights and needs. But regardless of society, a woman is always viewed through the prism of her procreative functions (de Beauvoir, 2010). This is especially true for the ruling elite, who have expressed and continue to express concerns about the overpopulation or underpopulation of their own country or the West, depending on the circumstances (Brown, 2019). Women, therefore, are a constantly vulnerable group due to their reproductive biology and the centrality of population politics. Since women's reproductive rights have been and continue to be limited and determined by the expectations of politicians and men in general, it is very likely that biomedical human enhancement involving gene editing will also be applied with population policy in mind in a way that does not take into account women's expectations and needs.

Planning for the Future in Space **207**

Women may also be marginalized in ways other than those related to reproduction for those forms of human enhancement that would be directed at modifying morality and cognitive functions. Since sexist stereotypes are strong in all societies, access to particular types of biomedical human enhancement may also be shaped by these stereotypes and ideas about what is considered appropriate for each gender. If women are associated with gentleness, caring and concern for others, and are more often associated than men with the group that chooses to consider family more important than career, then the forms of bio-modification offered or available to women may be those that will perpetuate these stereotypical functions and roles. Practical policy depends on whether human enhancement with effects on morality and cognitive functions is treated as a means to abolish gender differences or as a tool to maintain those differences. In the former case, women may be subjected to modifications aimed at extinguishing their needs for having children and producing habits and needs characteristic of men. This relates to the sexist and patriarchal division between roles and functions associated with men and those associated with women, since in this traditional model women are associated with childcare and family life, while men are associated with the public sphere. A policy of applying biomedical human enhancement affecting morality and cognitive functions would therefore aim to swap these roles between the genders, that is, to make men morally and cognitively interested in devoting themselves to the family sphere, and women equally morally and cognitively interested only in devoting themselves to the public sphere.

Such policies can also take another form, in which only women will be subject to modifications to make their preferences more like those of men. Such a biomedical human enhancement policy could resemble a campaign to increase women's participation in fields of study associated with men, primarily STEM. In contrast, this campaign does not have an equivalent that targets men and encourages their increased presence in feminized fields of study such as nursing or psychology (Mullen & Baker, 2018).

However, it is doubtful that human enhancement policies are being implemented to level the playing field and minimize gender differences with regard to the aforementioned division between the private and public spheres. As we mentioned, the focus of the policy is female reproductive biology and population issues. Therefore, the ruling elite will not allow such a population-politically risky equalization between men and women. Therefore, it cannot be ruled out that the more likely goal of human enhancement policies will be to reinforce gender inequality by influencing women's morality and cognitive functions.

Since sexist and patriarchal philosophy considers women as inferior beings to men, it can be assumed, for example, that the types of biomedical human enhancement that are supposed to improve cognitive function and intelligence will not be applied to women. The argument for excluding women may be that women who are considered inferior do not need to improve intelligence or other functions that are not related to their functions in the private sphere. Even today, without a human enhancement policy, in many parts of the world, girls are denied education precisely because of this patriarchal philosophy (Szocik, 2023).

We see that women can be excluded simply because they are women – that is, they are considered inferior, they are associated with jobs in the private sphere, as well as with reproduction. As long as this patriarchal philosophy persists, human enhancement will not serve women. On the contrary, it has the potential to reinforce the roles traditionally assigned to them.

Another vulnerable group that can be victimized by biomedical human enhancement policies are exploited workers under capitalism. There are many different reasons why people can work longer and more in capitalism. Human enhancement that increases productivity at work can be used for this purpose, to work longer and more. The effect of a social and political nature is to preserve the system of human exploitation, as employers will use the existence of a sizable group of workers undergoing human enhancement to put pressure on everyone else. It is also impossible to rule out a hypothetical scenario in which information about undergoing human enhancement for work or useful from the employer's point of view would be available to the employer and used in the recruitment of employees (Sparrow, 2019).

We believe that the vulnerable group at particular risk of exclusion and marginalization as a result of the introduction of human enhancement policies are people with disabilities. These people will find themselves under multiple oppression and will experience exclusion in the following ways. First, they will feel obligated to undergo biomedical human enhancement for corrective purposes, that is, to bring these people into line with the current ableist norm. People with disabilities will thus discipline themselves in an ableist culture (Scully, 2022). They will gain a new tool to bring themselves into line with the norm, so that they will feel less or no exclusion that they experience as people with disabilities. This phenomenon will be an example of self-discipline within the framework of bio-politics created based on biomedical human enhancement tools. Secondly, the concept of biomedical human enhancement applied to people with disabilities understood as public policy bears the characteristics of a vicious circle. The paradoxical nature of this policy is that the target group is supposed to be people whose performance and skills in a given area are lower than those of able-bodied persons. Thus, as long as able-bodied candidates are available on the market, improving the condition of a disabled person who is lower than that of an able-bodied person is devoid of justification from the point of view of the employer's interests and task. Only state entities that are guided by the ideals of equality and antidiscriminatory policies are capable of carrying out and monitoring such human enhancement policies toward people with disabilities. This second case of discriminating against people with disabilities as those with lower productivity may apply to those jobs where a certain high productivity is required. The employer may then prefer to invest in the enhancement of a candidate who already exhibits sufficiently high performance than to invest in biomedical procedures in a candidate whose performance is substandard and yet to be corrected.

This case is analogous to gender exclusion, where women may be excluded from biomedical human enhancement policies on the grounds that they are considered inferior to men in terms of specific characteristics relevant to employers in a given profession. This may be the case, for example, for soldiers,

Planning for the Future in Space **209**

where military service has traditionally been associated as the domain of men rather than women, where decision-makers may not see the point in enhancing women (which from their point of view will merely bring women up to the level of performance inherent in men), only to focus on enhancing men (which from the point of view of those with power will literally be true enhancement above the norm).

The policy of biomedical human enhancement is not innocent and neutral from a bioethical point of view. Its implications may be surprising if we pay attention to the potential impact its application may have on traditionally excluded and marginalized groups such as women, workers, and people with disabilities. At the same time, it shows that technologies, especially biomedical technologies, are not neutral.

Human Enhancement and Vulnerable People in Space

It can be surmised that if biomedical human enhancement is likely to strengthen the mechanisms of discrimination and exclusion with regard to the aforementioned groups on Earth, it will carry all the more risk of exclusion, oppression, and discrimination with regard to the cosmos. This is consistent with our intuitions inherent in the idea of moral exceptionalism. This does not mean that we agree with this differential treatment of people in space and the loosening of our moral requirements. It means that exclusion and discrimination will be very likely precisely because of the difficult conditions in space, which correlates with our intuitive readiness to accept exceptions to the norm in difficult circumstances. While we do not accept the idea of moral exceptionalism (see also Balistreri & Umbrello, 2022), it is worth noting the potential areas of risk of replicating exclusion and discrimination against those taking part in space missions in a situation where biomedical human enhancement would be used for these missions.

It seems that the same groups of people we have considered vulnerable with regard to human enhancement policies on Earth will face exclusion and discrimination in space. In the near future, the most vulnerable group will be employees of space corporations, but also of state space agencies – all those who will take part in long space missions to the Moon, Mars, or asteroids. Subjecting them to human enhancement will serve to protect their health and lives, as well as increase productivity. However, this goal, which is good from a moral point of view, cannot be isolated from the goal of space missions. These goals, as we have indicated, are morally wrong because they grow out of militarism, nationalism, capitalism, and colonialism. Thus, it can be said that a good means, namely health care, serves a bad end, namely nationalist and exploitative politics. But it is also the case that the wrong end is served by the wrong means, that is, strengthening worker productivity, which is itself exploitation.

Both in missions organized by state agencies, which will be nationalistic, militaristic, and economic in nature, as well as in missions organized by private agencies, which will be economic in nature, all participants subjected to human enhancement will be exploited to advance government and corporate interests. This is the most important element of future space policy that constitutes the nature of moral judgment. Workers on space missions will be a very vulnerable group due to environmental conditions, namely confinement in a small space, inability to migrate and possibly escape, as well as distance from Earth. In a sense, subjecting them to biomedical human enhancement, whether mandatory or optional, will not increase the difficulty of their vulnerable situation. Following Erika Nesvold, it is worth noting the particular exploitation and deprivation of fundamental rights of future employees of space corporations. Even if we consider it appropriate to guarantee labor rights familiar from Earth, it is not clear whether they can be enforced in space. It is also unclear what the financing system for transporting and maintaining a worker in space will be, given the high cost of space missions. Finally, it is also unclear to what extent working in space will be considered attractive. All of these circumstances will negatively affect the legal status, actual opportunities, and empowerment of workers in space (Nesvold, 2023a; 2023b). Human enhancement applied to workers will therefore be a tool to increase their exploitation by making them even more useful and efficient employees. It will also deepen their objectification and commodification in line with the view that high risk and cost missions require adequate preparation of workers. It can also be assumed that the availability of forms of biomedical human enhancement affecting morality and cognitive functions will be exploited in a manner analogous to the situation described in relation to gender. That is, just as from the point of view of capitalism and patriarchy a woman is viewed through the prism of her reproductive functions and responsibilities in the private sphere, the worker is viewed through the prism of capital accumulation. Since the greater the exploitation of the worker, the greater the accumulation of capital, the nationalist, and capitalist forces guiding space politics may find it reasonable to subject workers to appropriate moral and cognitive modifications. The effect of such modifications would be to increase workers' reconciliation with their fate.

Another vulnerable group in space, whose situation will be definitely worsened by human enhancement, are women. However, we are talking here about a specific scenario belonging to the distant future, in which humans will live in permanent or near-permanent bases in space and reproduce there. The possibility of human reproduction in space will make women the center of population politics, which will mean the restriction of women's reproductive rights, which will be subordinated to group interests. All biomedical technologies related to reproduction will lead to the instrumental treatment of women. A particular threat could be germline gene editing, which could act as a tool for social and political engineering at the space base. It would therefore be problematic not only for women themselves but also for this population in general, if its reproduction is designed according to the partisan vision of the mission organizers and sponsors (Kendal, 2023).

A group specifically under permanent oppression and excluded probably more than all other vulnerable groups are people with disabilities. Indeed, the logic of applying human enhancement with the idea of increasing adaptation to harsh conditions and enhancing performance excludes people with disabilities.

From the point of view of the interests of mission organizers, both public and private, people with disabilities are adapted to the conditions of space missions to a lesser degree than able bodied people. They therefore require intensified modification compared to able bodied candidates, because in addition to upgrading them to the level required for a given mission, they also require matching their level to the standard set by ableism. The context, that is, the harsh conditions in space and the costliness and risk of missions, reinforce the rationale for this logic, leading to a situation in which people with disabilities do not fit the image of space missions as something extremely difficult, dangerous, and highly specialized (Szocik, 2024).

Among vulnerable groups, we do not include racial minorities. If the main or only type of human missions in the future will be missions focused on the exploitation of space resources, private corporations as well as state agencies will need specialists. Racist biases will not necessarily come into play here. On the other hand, it is worth pointing out that the racist bias is global capitalism, which began its career precisely with racist slavery and labor exploitation. Today's global neoliberal capitalism also exploits the global South in particular (Mattig, 2023).

Conclusions

What should our future be in space – with or without radical biomedical human enhancement? From the standpoint of safety, health, and the impact of environmental conditions, we regard biomedical modifications for space missions as routine procedures as long as they are safe and effective. This is not a trivial requirement, as illustrated, for example, by the unpredictability of the effects of genetic modification depicted in the example of the main protagonist of the film *The Titan*, Lt. Rick Janssen, played by Sam Worthington. In order to adapt future astronauts to the conditions on Titan, Janssen and others are subjected to experimental genetic modification, which proceeds in an uncontrolled and unpredictable manner. Only Janssen, heavily modified, survives; the others die during the modification procedures or are shot by the military (The Titan, 2018). Admittedly, the modifications we are talking about are not as radical as those depicted in the film, the purpose of which was to allow the modified astronauts to live on the surface of Titan without life support system. However, the film does a good job of showing the risks and unpredictability of gene editing from a medical and biological point of view. It also does an excellent job of showing the astronaut's family context, the shock experienced by the wife upon meeting her genetically modified husband. These are undoubtedly important contexts, including for feminist bioethics, which pays attention to the relationality of the individual. However, issues of safety and efficacy are the ethical minimum. Even after they are met, the autonomy, freedom, and rights of modified individuals are not guaranteed. Thus, if we pay attention to the context of autonomy, equality, and individual rights, our assessment of the application of biomedical human enhancements to future space missions is negative. The environment of future space missions will exploit everyone, while some traditionally vulnerable and

212 Rakhat Abylkasymova and Konrad Szocik

excluded groups such as women, people with disabilities and workers will be under particular oppression. Under conditions of the Capitalocene (Federau, 2023) and the Great Acceleration (McNeill, 2023), any project, especially one as costly and risky as space exploration and exploitation, must fit into the logic of capitalist exploitation. Human enhancement will reinforce these mechanisms. Evaluation of biomedical procedures should not be done independently of the purpose and rationale for space missions.

Acknowledgments

(Konrad Szocik) This work was supported by the National Science Centre, Poland (UMO 2021/41/B/HS1/00223).

References

Balistreri, M., & Umbrello, S. (2022). Space travel does not constitute a condition of moral exceptionality. That which obtains in space obtains also on Earth. *Medicina e Morale, 71*(3), 311–321.

Ballantyne, A. (2022). Women in research. Historical exclusion, current challenges and future trends. In W. A. Rogers, C. Mills, & J. Leach Scully (Eds.), *Routledge handbook of feminist bioethics* (pp. 251–264). Routledge.

Brown, J. (2019). *Birth Strike: The Hidden Fight over Women's Work*. PM Press.

de Beauvoir, S. (2010). *The second sex*. Translated by Constance Borde and Sheila Malovany-Chevallier; with an introduction by Judith Thurman. Alfred A. Knopf.

Federau, A. (2023). Capitalocene. In N. Wallenhorst & C. Wulf (Eds.), *Handbook of the anthropocene* (Vol. 1, pp. 641–644). Springer.

Garasic, M. D. (2021). The war of ethical worlds: Why an acceptance of Post-humanism on Mars does not imply a follow up on Earth. *Medicina e Morale, 70*(3), 317–327.

Gregg, B. (2023). Genetic engineering revolution. In N. Wallenhorst & C. Wulf (Eds.), *Handbook of the anthropocene* (Vol. 1, pp. 505–510). Springer.

Hauskeller, M. (2024). Foreword. In N. Agar (Ed.), *Dialogues on human enhancement* (pp. ix–xi). Routledge.

Kendal, E. (2023). Desire, duty, and discrimination: Is there an ethical way to select humans for Noah's Ark? In J. S. J. Schwartz, L. Billings, & E. Nesvold (Eds.), *Reclaiming space: Progressive and multicultural visions of space exploration* (pp. 289–302). Oxford University Press.

Mattig, R. (2023). Racism. In N. Wallenhorst & C. Wulf (Eds.), *Handbook of the anthropocene* (Vol. 2, pp. 1535–1540). Springer.

McNeill, J. R. (2023). Great Acceleration. In N. Wallenhorst & C. Wulf (Eds.), *Handbook of the Anthropocene* (Vol. 1, pp. 821–824). Springer.

Mullen, A. L., & Baker, J. (2018). Gender gaps in undergraduate fields of study: Do college characteristics matter? *Socius, 4*. https://doi.org/10.1177/2378023118789566

Nesvold, E. (2023a). *Off-Earth: Ethical questions and quandaries for living in outer space*. The MIT Press.

Nesvold, E. (2023b). Protecting labor rights in space. In J. S. J. Schwartz, L. Billings, & E. Nesvold (Eds.), *Reclaiming space: Progressive and multicultural visions of space exploration* (pp. 241–250). Oxford University Press.

Rubenstein, M. (2022). *Astrotopia: the dangerous religion of the corporate space race.* The University of Chicago Press.

Russell, C. (2022). What makes an antiracist feminist bioethics? In W. A. Rogers, J. Leach Scully, S. M. Carter, V. A. Entwistle, & C. Mills (Eds.), *The Routledge handbook of feminist bioethics* (pp. 195–207). Routledge, Taylor & Francis Group.

Schwartz, J. S. J., & Milligan, T. (Eds.) (2016), *The ethics of space exploration.* Springer.

Scully, J. L. (2022). Feminist bioethics and disability. In W. A. Rogers, C. Mills, & J. Leach Scully (Eds.), *Routledge handbook of feminist bioethics* (pp. 181–194). Routledge.

Sparrow, R. (2019). Yesterday's child: How gene editing for enhancement will produce obsolescence – And why it matters. *The American Journal of Bioethics, 19*(7), 6–15.

Szocik, K. (2023). Cognitive enhancement inevitably leads to discrimination against women. *AJOB Neuroscience, 14*(4), 357–359.

Szocik, K. (2024). *Feminist bioethics in space.* Oxford University Press.

Szocik, K., & Braddock, M. (2019). Why human enhancement is necessary for successful human deep-space missions. *The New Bioethics, 25*(4), 295–317.

The Titan. (2018). *The Titan.* IMDb. https://www.imdb.com/title/tt4986098/

Chapter 11

Building Better (Space) Babies: Bioastronautics, Bioethics, and Off-World Ectogenesis

Evie Kendal

Swinburne University of Technology, Australia

Abstract

One of the more plausible methods of establishing an off-world human society is to transport a small number of adults alongside a large supply of frozen embryos. This would minimize the costs of transportation and protect genomic diversity across future generations. However, there are various ethical issues with attempting off-world pregnancies, in terms of unknown risks to pregnant people and fetuses, and potential discrimination concerns if the adults selected for the mission are required to be willing and able to gestate this genetically diverse population of embryos. An alternative would be the development and use of artificial womb technology (ectogenesis) to perform this latter function. Benefits might include freeing all crewmembers to devote all their energies to establishing the off-world facility without pregnancy-related illness affecting health and productivity, and providing a safer environment for fetal development, for example, providing additional radiation shielding around the artificial wombs. This chapter will explore some of the ethical issues surrounding ectogenesis and its space applications.

Keywords: Bioethics; bioastronautics; reproductive technologies; ectogenesis; artificial womb; space medicine

Introduction

With increasing government and commercial activities in space, and the prediction of a boom in space tourism in the future, new ethical issues are arising regarding human health and safety in this sector (Cohen & Spector, 2020).

The Ethics Gap in the Engineering of the Future, 215–227
Copyright © 2025 Evie Kendal
Published under exclusive licence by Emerald Publishing Limited
doi:10.1108/978-1-83797-635-520241012

216   Evie Kendal

Patel et al. (2023) note a major milestone in "public-private partnership" for space exploration occurred in 2020, with the first successful launch of astronauts into low-Earth orbit (LEO) on a commercial spacecraft. The number of private companies with space aspirations is ever-increasing, while the cost barrier to reach orbit continues to decline, providing access to what was once prohibitively expensive for all but the largest national space agencies. As Matsumura et al. (2019) claim: "The era where people can easily go into space is coming." Despite this growth, there is currently a lack of ethico-legal guidance for space exploration, particularly for private enterprises, and an identified need to establish human rights standards beyond our planet (Cockell, 2022).

This chapter focuses on one under-researched aspect of human off-world exploration: the need to develop a reproductive bioethics framework for space. This includes addressing ethical concerns about pregnancy, fetal development, and reproductive technologies relevant to space research, tourism, and long-term settlement, as well as other social elements of reproduction, such as employment rights and protecting against discrimination. The chapter will open by discussing how space agencies have handled the issue of reproductive health in space, before considering some of the potential risks for off-world pregnancies and how emerging technologies, such as ectogenesis, may contribute to fetal and astronaut well-being and promote human survival on another planet.

Historical Perspectives on Reproduction in Space: Institutional Sexism and the Positioning of Women's Bodies as the "Problem"

While some national space agencies have policies explicitly banning astronaut pregnancies due to the potential health risks to astronauts and fetuses, public speculation regarding sex in space and its possible consequences remain (Fessenden, 2015). As Antarctic programs – often considered analogues for space missions – demonstrate, strong sanctions and high risk are often insufficient impediments to fraternization, with several documented pregnancies occurring among staff stationed at remote Antarctic posts (and these in the absence of any additional draw a zero-gravity situation might inspire) (Wood et al., 2005). Growing interest in private space tourism and future off-world human settlements also demonstrate a need to urgently address the issue of reproduction in space, including its ethical dimensions.

Historically, the topic of sex in space has been actively avoided by organizations like NASA, as demonstrated by the following quote from Casper and Moore (1995), regarding astronauts Jan Davis and Mark Lee:

> In 1992, a married couple flew together on a U.S. space shuttle mission, generating a flurry of public curiosity and controversy over what the paparazzi termed 'celestial intimacy'. The National Aeronautics and Space Administration (NASA) was bombarded with questions about heterosexual sex and reproduction in space, topics which the agency seemed ill-equipped and unwilling to address.

> Not only are sex and reproduction perceived as topics which should not be discussed in polite society, they are also seen as contributing to a loss of legitimacy for NASA in an age of uncertain and ever-diminishing resources (p. 312).

These authors also note NASA's response to reproductive health issues in space has often been to consider male bodies "the norm," while female bodies are "configured as problematic" and in need of manipulation so they can "fit into the space program" (p. 312, 317). Pregnancy has been conflated with sexual activity in this heterosexist paradigm, with the supposed need to avoid the former leading to the various sanctions against the latter. However, Casper and Moore (1995) claim this had the effect of historically limiting research on sex-based differences and reproductive health for astronauts to just focusing on contraception for female astronauts, ignoring the impact of exposure to the space environment on sperm production and quality (p. 318). More recent studies have certainly started to challenge the assumption that space pregnancies are predominantly a women's issue, with numerous animal experiments focused on male reproductive function following space travel (Matsumura et al., 2019). This aligns with Earth-bound reproductive health research that suggests most embryo DNA damage is acquired from the male gamete, since mature sperm are far less resistant to genetic injury than the human oocyte (Ellegren, 2007; Furness et al., 2011).

When it comes to long-term settlement plans, banning copulation and off-world pregnancies becomes not only more practically challenging but also counter-productive. One of the more plausible methods of establishing an off-world human society would be to transport a small number of adults alongside a large supply of frozen embryos. This would minimize the costs of transportation and protect genomic diversity across future generations. However, there are various ethical issues with attempting off-world pregnancies, in terms of unknown risks to pregnant people and fetuses, and potential discrimination concerns if the adults selected for the mission are required to be willing and able to gestate this genetically diverse population of embryos. The freedom to choose if, when and with whom to procreate is often afforded significant moral weight in human societies, and it is also likely that some new settlers will prefer to pursue pregnancy through more traditional routes, even if this diminishes genetic diversity. In both cases, safety considerations are of paramount ethical importance.

While the former assumption that women's reproductive potential needed to be suppressed in the space environment should be criticised, it is nevertheless the case that being pregnant is space is likely to carry significant risks. Some of these represent an exacerbation of the risks pregnant people face on Earth, while others are unique to the space environment, as will be discussed below.

218 Evie Kendal

Off-World Pregnancy Risks

Living in microgravity and being exposed to high levels of cosmic radiation are known health hazards (Szocik, 2020); however, there has been no data collected to date on the effects of these phenomena on a real human space pregnancy. Before such a thing is attempted, we need to consider the potential health risks associated with this and map the ethico-legal landscape. The humorous anecdote about NASA allegedly asking Sally Ride, the first American woman in space, whether she would need 100 tampons for a 6-day mission, seems to indicate we are not ready for the complexities of managing reproductive health in space. Nevertheless, with longer-term space missions on the horizon, including in the private sector, such conversations need to be had. NASA's "Five hazards of human spaceflight" provides a useful starting point (NASA, 2018), with each point here specifically adapted to the hypothetical space pregnancy scenario:

Hazard 1: Space Radiation

Cosmic radiation consists of high-energy particles, X-rays, and gamma rays. Without the protection of Earth's atmosphere, astronauts are exposed to high levels of these damaging forms of radiation every day, which has been demonstrated to cause health problems, including an increased lifetime cancer risk (Hassan et al., 2020). Organ damage, malignancies, and reproductive problems have all been associated with prolonged exposure to cosmic radiation, with Hassan et al. (2020) also noting specific threats to pregnancies, including miscarriages and birth defects (p. 487). Because these forms of radiation can penetrate the skin and organs, DNA damage is likely to occur for both the pregnant person and the developing fetus, with Watson (2012) suggesting genomic stability within the oocyte may be negatively impacted by accumulated radiation dose even before fertilization can occur. Ionizing radiation causes oxidative stress and the production of free radicals, causing excessive DNA single and double-strand breaks, which in turn can lead to toxic and mutagenic lesions (Salmon et al., 2004). Kumar and De Jesus (2023) note that for the developing fetus, especially in its earliest stages, exposure to ionizing radiation can be fatal, or lead to malformation, growth retardation, mutations, and cancers. In general, the lower the gestational age of the fetus, the more profound the effects of radiation damage, with the first few weeks of embryogenesis being the most fragile time (Shaw et al., 2011). Some animal studies have indicated radiation damage can occur in frozen sperm samples exposed to the space environment, while others have shown no significant loss of reproductive function (Matsumura et al., 2019).

Another form of radiation astronauts are exposed to in large doses is ultra-violet (UV) radiation. In their dermatological study, "Celestial effects on the skin," Hassan et al. (2020) note UVA and UVB have various important functions, with both promoting melanin synthesis and UVB also being involved in vitamin D synthesis, through the initial conversion of 7-dehydrocholesterol in the skin to previtamin D3. This previtamin is then further processed in the liver to 25-hydroxyvitamin D3 and then finally in the kidney to its active form, 1,25-

dehydroxyvitamin D3 which has a role in promoting bone health (p. 486). However, these authors also note that excess UVA causes oxidative stress and free radical production, while UVB causes DNA damage through the formation of pyrimidine dimers (intra-strand cross-links between adjacent nucleotides that deform the helix structure and block DNA replication and transcription). The larger the dose of UV exposure, the greater the damage sustained, causing mutations and reducing tumor suppression activity. Excess UV exposure is also associated with infertility, particularly in males (Rajput et al., 2022).

At present, radiation risks are mitigated for astronauts through spacecraft shielding, with some further protection offered by their spacesuits. However, developing suits that can be adapted for rapid body changes may be challenging, so future pregnant space travelers may find their movements are more restricted than other crewmembers due to their increased vulnerability, and that of their fetus. Being limited to duties that can be conducted in the most heavily shielded areas of space vessels may exacerbate the next threat to be considered, the effects of isolation.

Hazard 2: Isolation and Confinement

Isolation itself is unlikely to impact fetal development directly; however, the psychological strain of prolonged isolation is likely to be exacerbated by the hormonal changes occurring in pregnancy. The lockdowns many countries initiated during the COVID-19 pandemic caused significant mental health issues among certain populations, notably including pregnant individuals who reported higher levels of depression and anxiety caused by increased stress and the need to socially distance from their support networks (Kuipers et al., 2022). Translating this to the space environment, we can expect to see similar patterns of stress among pregnant astronauts who will have even fewer supports available. Casper and Moore (1995) even suggest a pregnant crewmember may experience targeted social exclusion by teammates who resent any reduced work capacity or contribution to the mission caused by pregnancy-related illness and later childrearing duties (p. 326). They also note a pregnant astronaut may have different environmental and nutritional needs than the other crewmembers, leading to potential conflict when living in close quarters and with limited supplies. In terms of risks to the fetus, untreated depression in pregnant women has been associated with reduced alertness and developmental delays in offspring (Jahan et al., 2021). Prenatal depression is also a risk factor for postnatal depression and other psychological sequelae (Jahan et al., 2021). If the childbirth is traumatic, the risks increase further, which is itself more likely due to the isolation of space travel and distance from Earth.

Hazard 3: Distance From Earth

The most obvious threat to a pregnant person's survival in space refers to the limited capacity for surgical interventions if required, for example, an emergency Caesarean birth. Other complications include the need to deliver prenatal medical

220   Evie Kendal

care, dietary supplies, and other pregnancy-related needs and to have a skilled birth attendant present to ensure a safe delivery. Many obstetric emergencies can lead to rapid deterioration of health, with survival chances often being directly related to the distance to a tertiary care facility (Scott et al., 2013). Astronauts will be further from hospital than any other pregnant individual, thereby increasing the risk of maternal and fetal/neonatal mortality.

Hazard 4: Gravity Fields

Another major health hazard for humans in space is exposure to microgravity and the different gravity conditions of other celestial bodies. While earlier predictions that female astronauts might experience "retrograde menstruation" if weightless have proven false, many biological systems on Earth are dependent on gravity, including some occurring during pregnancy. It is also noteworthy that various tips for easing childbirth or managing lactation issues also rely on gravity, for example, adopting upright or squatting birthing positions, or reducing excessive milk flow by lying down following the let-down reflex. Yamasaki et al. (2004) also note that gravity is responsible for producing the hydrostatic pressure that influences blood pressure gradients, with the strongest variation occurring when someone stands upright under normal Earth gravity conditions (p. 819). These authors further explain that microgravity causes body fluids to shift toward the head due to the removal of this pressure gradient, ultimately reducing blood volume and pressure throughout the body (p. 820). In their animal experiments, they found that neonatal rats exposed to microgravity on the space shuttle did not grow as large as normal rats, summarizing their results as follows:

> The rat neonates, at least surviving ones, were healthy in space but were smaller, suggesting that body weight gain would be suppressed mainly in accordance with the reduction of body fluid volume due to its headward shift and redistribution. A reduction of the mass of skeletal muscles and/or bone due to hypokinesia under weightlessness would also enhance the body weight suppression (p. 826).

Loss of skeletal muscle and bone demineralization are also known to affect adult astronauts exposed to microgravity (Szocik, 2020), and there are significant concerns regarding the increased impact expected for growing children. There are no data available on how microgravity might impact fetal development of muscles or bone, or whether a fetus will know how to move itself into a head-down position to facilitate birth without the influence of gravity. Patel et al. (2023) note that 20 years of data from humans on the International Space Station have indicated that variations to the gravity conditions humans are used to significantly disrupts homeostasis, as our "cardiovascular architecture and regulatory mechanisms have evolved to respond directly to the earth's force of gravity and sustain

Building Better (Space) Babies **221**

homeostasis via hemodynamic modulation." These authors also note similar adaptations in the musculoskeletal system, with bones and muscles being strengthened due to the resistive force of Earth's gravity acting on them, and exposure to the space environment therefore causing "atrophy, diminished force, and loss of functionality." Preventive measures for astronauts include targeted exercise regimes that a pregnant crewmember may experience greater difficulty completing. Pregnancy is also already associated with mineral loss for many due to the increased nutrient demands of the fetus, thus increasing the overall risks (Kovacs, 2016).

Patel et al. (2023) also outline various anatomical structures that are impacted by microgravity, including those in the cardiovascular, digestive, and reproductive systems. For the latter, they focus on the morphological changes that occur in the mammary gland of a pregnant individual to prepare for lactation, claiming exposure to the space environment impacts the genes responsible for managing this and other physical transitions. Matsumura et al. (2019) add disorders of the optic nerve to the list of health hazards associated with microgravity, as well as reminding us that hypergravity during launch and landing maneuvers also represent a threat to human safety, due to decreases in "cerebral blood flow and arterial pressure." Circulatory system adaptations made to sustain a pregnancy mean pregnant crewmembers will be at higher risk of adverse events resulting from other stressors on this system, including reduced oxygen, higher altitude, and lower pressures.

Hazard 5: Hostile/Closed Environments

As humans are not adapted to live in space, all their survival needs must be met within the confines of their spacecraft. Safety concerns include managing microbes within vessels, and ensuring appropriate temperatures, air quality, lighting, and noise levels are maintained (NASA, 2018). Space travel has been shown to alter immune function, with some astronauts experiencing reactivation of certain viruses, such as herpes (Akiyama et al., 2020). Immune function is also impaired by sleep disturbances, which are a likely result of alterations to natural lighting and astronauts' circadian rhythms while off-world (Hassan et al., 2020). Pregnancy is also a special immune state, allowing the body to tolerate the presence of a foreign entity (the fetus) without the immune system attacking it. However, it is not yet known whether immune system alterations in space will undermine this process, potentially risking the health of pregnant astronauts or their fetuses.

Other concerns regarding the closed systems used in space travel and how they intersect with reproductive health include what Casper and Moore (1995) refer to as challenges to "cosmic abstinence," sexual privacy, and the threat of sexual violence (p. 322, 325). Sexually transmitted infections, potentially including ones caused by the reactivation of viruses, also pose a health risk. There are also concerns that if children are grown and develop off-world, the effects of maturing under different radiation and gravity conditions and other environmental factors

222 *Evie Kendal*

will mean they are no longer adapted to live on Earth, should they wish to return (Casper & Moore, 1995). Other issues directly relate to gender-based discrimination in reproduction, as will be discussed in the next section.

Overall, the potential health hazards of space pregnancy are many and varied, but one way of mitigating them would be to resituate gestation outside the human body using technology. While this solution may not be embraced by all future off-world settlers, using an artificial womb (ectogenesis) to help populate a new off-world settlement could reduce some of the physical, psychological, and social risks of reproduction that would otherwise disproportionately impact women.

Potential Benefits of Using Ectogenesis for Space Settlements

Casper and Moore (1995) claim that despite the systematic exclusion of women from space programs and concerns about their reproductive potential, "women astronauts are defined simultaneously as potential sexual partners for male astronauts and as potential reproducers in the interest of colonization" (p. 312). These authors further claim that plans for off-world human settlements represent "the realization of penetration and colonization fantasies about the future," suggesting space exploration has been a "historically masculine project" as exemplified by the phallic designs of most rocket models (p. 316). This leads to concerns regarding women's participation in this colonizing effort, with gestational ability potentially becoming a commodity for which women's bodies could be trafficked, as has happened throughout colonial history (p. 327). Returning to the idea of transporting large numbers of embryos to found a new off-world human society, the need for women's bodies to gestate the next generation of space travelers might severely limit their other opportunities, leading to a form of reproductive slavery (Kendal, 2023). One method to avoid this would be to ensure – at least the option of – artificial gestation.

Ectogenesis refers to the development of a fetus outside of the organic womb and the potential for this technology has fascinated scientists since at least the 1950s. Prototype artificial wombs are already in development around the world, including the Extrauterine Environment for Neonatal Development (EXTEND) therapy system, designed by the Children's Hospital of Philadelphia, United States, and the Ex-vivo Uterine Environment (EVE) platform, representing a collaboration between the University of Western Australia and Tohoku University Hospital in Japan (Romanis & Kendal, 2024). Both of these models involve a fluid-filled biobag in which premature lamb fetuses have been successfully maintained for between one and four weeks. In these and other ectogenesis models, gas and nutrient exchange has been managed through a series of arterio-venous pumps, external membrane oxygenators, and/or umbilical catheters (Mercurio, 2018). A prototype for use in human trials is currently under development at the Eindhoven University of Technology in the Netherlands, alongside projects focused on devising methods of transferring a partially developed fetus from the organic womb to the artificial womb. Current research in the area is focused on improving survival rates for infants born extremely prematurely, but long-term applications might include full ectogenesis (gestation from conception to birth). This latter possibility yields the most promise for use in off-world settlements.

The social benefits of providing an alternative site for human gestation include removing certain forms of pregnancy-based discrimination and the potential that crew selection will be unduly influenced by gestational capability. It would also ensure female settlers are not subjected to undue pressure to provide reproductive services to advance the settlement, while also protecting their productivity by eliminating pregnancy-based illness and reduced function. These elements promote both individual autonomy and the good of the collective.

In terms of mitigating the hazards of spaceflight as pertaining to pregnancy, an ectogenesis chamber could be protected by levels of radiation shielding not achievable in the in vivo environment, at least not without severely limiting the freedom of movement of the pregnant crewmember. Expecting a child would not need to alter any crewmembers' allocated tasks, even if required to work in an unshielded section of a craft, as the fetus would be safe elsewhere. This would also minimize any additional psychological risks associated with isolation and confinement due to pregnancy. Being far from the Earth would continue to pose a risk to fetal well-being; however, removing the potential need for surgical delivery or possibility of obstetric emergencies would significantly reduce the risk to crewmembers who would otherwise be giving birth off-world. Perhaps most excitingly, an ectogenesis chamber could be maintained within an artificial gravity system to avoid any health risks associated with fetal exposure to microgravity. Matsumura et al.'s (2019) mouse fertility study developed hardware that generated gravity at approximately Earth normal values using centrifugal motion, and successfully demonstrated that mice maintained in such a system had similar reproductive outcomes as mice living on the ground. Finally, nutrition and oxygenation could be optimally calibrated in an external gestational system with greater ease than in the biological system, removing the need to substantially alter the living conditions of the crew to meet the needs of a pregnant crewmember. Exposure to potentially harmful microbes could also be avoided without imposing on other settlers. The artificial womb would essentially be a closed system within the closed system of the spacecraft or off-world habitation, improving the safety of the developing fetus within the context of the hostile space environment.

A Reproductive Bioethics Framework for Space

What remains is the need to establish a systematic ethical approach to human reproduction in space, taking into account the potential role of emerging technologies for supporting future human settlements. Ectogenesis has not yet been successfully demonstrated, so part of the reproductive bioethics framework being proposed here relates to the ethical testing of this unproven technology. This is a particular concern when considering the core principles of research ethics include respecting the autonomy of research participants and ensuring they give informed consent for experiments.

224   Evie Kendal

Many areas of bioethics involve uncertainty, not only because they focus on future technologies where risks and benefits may be unclear but also because they must consider how current practices in the biosciences and medicine can and should be changed for ethical reasons. One of the greatest ethical challenges facing reproductive medicine arises because new technologies necessarily carry unknown risks to women and their future offspring. This calls into question the possibility of providing truly informed consent for such novel interventions. When translated to the harsh and remote environment of space, these risks become even more uncertain.

Informed consent is the cornerstone of medical ethics and typically requires that medical professionals explain pertinent details of proposed interventions to clients, thereby equipping them to make decisions regarding their health and medical care. To engage in a rational assessment of personal risks and benefits, individuals need to be able to exchange information with those possessing knowledge of their risk profile and the relevant threats (Reynolds & Seeger, 2005). This process, known as risk communication, relies on health professionals having access to specialized knowledge; however, in the case of novel technologies or applications, relevant evidence might not be available when needed. For potential parents considering ectogenesis, knowledge about its success rates and risks is still unknown, meaning the early adopters will be engaged in clinical experimentation.

When discussing concerns about fetal experimentation, Singer and Wells (2006) note that if it would be considered unethical to experiment with ectogenesis technology until we have "a reasonable assurance that it is safe and we can have no reasonable assurance that it is safe until it has been carried out, we seem to be in a classic 'catch 22' situation. Work on ectogenesis will remain forever unjustifiable" (p. 22).

The rationale behind this argument is that we know organic wombs are capable of gestating fetuses safely, so deliberately exposing a fetus to an unproven alternative is introducing unnecessary risk. That this risk would be imposed on a nonconsenting future person who would potentially have to live with any adverse effects of the technology, further complicates the matter. While current prototypes have only been trialled with animal fetuses, Singer and Wells (2006) argue that the above safety concerns mean human trials will likely depend on "serendipitous" extremely premature births to provide test subjects. As these infants presently have very poor prognoses, the risk threshold justifying exposing them to an experimental ectogenesis technology is much lower than if experimenting with artificially gestating offspring from scratch, for example, using IVF embryos. However, the current thought experiment considers emerging artificial gestation technologies in a situation where there is no established, safe status quo to compare it to. It is possible ectogenesis will prove to be safer than biological pregnancy in the space context; it would allow more control over nutrition in a situation where it is expected there will be very limited dietary options available, as well as avoid the need for pregnant crewmembers to restrict their duties due to concerns about radiation and the effects of microgravity on fetal development. Considering the ethical implications of ectogenesis in space thus opens up the possibility of evaluating the technology's merits divorced from assumptions regarding its comparable safety and integration into existing healthcare systems on Earth.

There are also legal considerations to address, including that in some jurisdictions around the world an infant "born" from an artificial womb would technically be considered parentless according to current laws (Bayne & Kolers, 2003). In the space environment, issues of citizenship and parentage may become even more complex. Changing such laws will require both scientific and ethical justifications. According to Philip A.E. Brey (2012), "[t]he ethics of emerging technology is the study of ethical issues at the R&D and introduction stage of technology development through anticipation of possible future devices, applications, and social consequences" (pp. 3–4). Bioethicists are already anticipating the expansion of current prototypes of ectogenesis technology, and suggestions this technology might be useful for space exploration and future human settlements are already being made (Braddock, 2018; Edwards, 2021).

Regarding expanding the field of bioethics more generally to engage reproductive bioastronautics, considering emerging reproductive technologies in the partially abstracted space environment allows ethical considerations to be evaluated separate to the integration of the technologies into existing social structures on Earth. The ethical justifications for developing these reproductive biotechnologies are also unique when considering the space environment is one that is hostile to human survival, potentially providing stronger moral reasons to pursue technologies to enhance survival advantage. In all clinical experimentation risks and benefits have to be weighed to determine the best way forward, but the unique health hazards of space exploration suggest a concentrated effort to mitigate these threats needs to be engaged, including in reproduction.

Conclusion

The current state of play in public and private space enterprises indicates an ever-increasing need to explore and understand what can and will be possible in medical science for promoting future human survival in space. With regards to mitigating the possible physical, psychological, and social hazards of biological reproduction in space, developing ectogenesis technology represents a significant step toward a viable off-world settlement strategy that promotes reproductive liberty and human health.

226 Evie Kendal

References

Akiyama, T., Horie, K., Hinoi, E., Hiraiwa, M., Kato, A., Maekawa, Y., Takahashi, A., & Furukawa, S. (2020). How does spaceflight affect the acquired immune system? *NPJ Microgravity, 6,* 14.

Bayne, T., & Kolers, A. (2003). Towards a pluralist account of parenthood. *Bioethics, 17*(3), 221–242.

Braddock, M. (2018). Next steps in space travel and colonization: Terraforming, ectogenesis, nano spacecraft and avatars. *Significances of Bioengineering & Biosciences, 2*(4), 167–173.

Brey, P. A. E. (2012). Anticipatory ethics for emerging technologies. *NanoEthics, 6,* 1–13.

Casper, M. J., & Moore, L. J. (1995). Inscribing bodies, inscribing the future: Gender, sex, and reproduction in space. *Sociological Perspectives, 38*(2), 311–333.

Cockell, C. S. (2022). *Interplanetary liberty: Building free societies in the cosmos.* Oxford University Press.

Cohen, E., & Spector, S. (2020). Space tourism – Past to future: A perspective article. *Tourism Review, 75*(1), 136–139.

Edwards, M. R. (2021). Android NOAHS and embryo arks: Ectogenesis in global catastrophe survival and space colonization. *International Journal of Astrobiology, 20*(2), 150–158.

Ellegren, H. (2007). Characteristics, causes and evolutionary consequences of male-biased mutation. *Proceedings of the Royal Society B: Biological Sciences, 274*(1606), 1–10.

Fessenden, M. (2015, April 2). Houston, we might have some major problems making babies in space. *Smithsonian Magazine.* https://www.smithsonianmag.com/smart-news/houston-we-might-have-some-major-problems-making-babies-space-180954828/

Furness, D. L. F., Dekker, G. A., & Roberts, C. T. (2011). DNA damage and health in pregnancy. *Journal of Reproductive Immunology, 89*(2), 153–162.

Hassan, S., Bhatti, J., Poulos, C., Mahmoud, A., Mohammed, T. O., & Hoenig, L. J. (2020). Celestial effects on the skin. *Clinics in Dermatology, 38,* 485–488.

Jahan, N., Went, T. R., Sultan, W., Sapkota, A., Khurshid, H., Qureshi, I. A., Alfonso, M. (2021). Untreated depression during depression and its effect on pregnancy outcomes: A systematic review. *Cureus, 13*(8), e17251.

Kendal, E. (2023). Desire, duty, and discrimination: Is there an ethical way to select humans for Noah's Ark? In J. S. J. Schwartz, L. Billings, & E. Nesvold (Eds.), *Reclaiming space: Progressive and multicultural visions of space exploration* (pp. 289–302). Oxford University Press.

Kovacs, C. S. (2016). Maternal mineral and bone metabolism during pregnancy, lactation, and post-weaning recovery. *Physiological Reviews, 96*(2), 449–547.

Kuipers, Y. J., Bleijenbergh, R., Van Den Branden, L., van Gils, Y., Rimaux, S., Brosens, C., Claerbout, A., & Mestdagh, E., (2022). Psychological health of pregnant and postpartum women before and during the COVID-19 Pandemic. *PLoS One, 17*(4), e0267042.

Kumar, R., & De Jesus, O. (2023, August 23). *Radiation effects on the fetus.* StatsPearls. https://www.ncbi.nlm.nih.gov/books/NBK564358/

Matsumura, T., Noda, T., Muratani, M., Okada, R., Yamane, M., Isotani, A., Kudo, T., Takahashi, S., & Ikawa, M. (2019). Male mice, caged in the International Space Station for 35 days, sire healthy offspring. *Scientific Reports, 9*, 13733.

Mercurio, M. (2018). The EXTEND system for extrauterine support of extremely premature neonates: Opportunity and caution. *Pediatric Research, 84*, 795–796.

National Aeronautics and Space Administration (NASA). (2018). Five hazards of human spaceflight. https://www.nasa.gov/stem-content/five-hazards-of-human-spaceflight/

Patel, O. V., Partridge, C., & Plaut, K. (2023). Space environment impacts homeostasis: Exposure to spaceflight alters mammary gland transportome genes. *Biomolecules, 13*, 872.

Rajput, N., Thakur, M., Highland, H., & George, L.-B. (2022). Deleterious impact of short duration UV-A exposure on the human sperm cell –An *in vitro* study. *Journal of Photochemistry and Photobiology A, 9*, 100093.

Reynolds, B., & Seeger, M. W. (2005). Crisis and emergency risk communication as an integrative model. *Journal of Health Communication, 10*(1), 43–55.

Romanis, C., & Kendal, E. (2024). Subjective experience, gestational preferences and justice: Valuing both uterus transplantation and ectogestation. In N. Hammond-Browning & N. J. Williams (Eds.), *International legal and ethical perspectives on uterine transplantation* (pp. 104–123). Edward Elgar Publishing.

Salmon, T. B., Evert, B. A., Song, B., & Doetsch, P. W. (2004). Biological consequences of oxidative stress-induced DNA damage in *Saccharomyces cerevisiae*. *Nucleic Acids Research, 32*(12), 3712–3723.

Scott, S., Chowdhury, M. E., Pambudi, E. S., Qomariyah, S. N., & Ronsmans, C. (2013). Maternal mortality, birth with a health professional and distance to obstetric care in Indonesia and Bangladesh. *Tropical Medicine and International Health, 18*, 1193–1201.

Shaw, P., Duncan, A., Vouyouka, A., & Ozsvath, K. (2011). Radiation exposure and pregnancy. *Journal of Vascular Surgery, 53*(1), 28S–34S.

Singer, P., & Wells, D. (2006). Ectogenesis. In S. Gelfand & J. Shook (Eds.), *Ectogenesis: Artificial womb technology and the future of human reproduction* (pp. 9–25). Rodopi.

Szocik, K. (2020). Human future in space and gene editing: Waiting for feminist space ethics and feminist space philosophy. *Theology and Science, 18*(1), 7–10.

Watson, A. J. (2012). Outer space and oocyte developmental competence. *Biology of Reproduction, 86*(3), 75.

Wood, J., Schmidt, L., Lugg, D., Ayton, J., Phillips, T., & Shepanek, M. (2005). Life, survival, and behavioral health in small closed communities: 10 years of studying isolated Antarctic groups. *Aviation Space & Environmental Medicine, 76*(Suppl. 6), B89–B93.

Yamasaki, M., Shimizu, T., Katahira, K., Waki, H., Nagayama, T., O-Ishi, H., Katsuda, S., Miyake, M., Miyamoto, Y., Wago, H., Okouchi, T., & Matsumoto, S. (2004). Spaceflight alters the fiber composition of the aortic nerve in the developing rat. *Neuroscience, 128*, 819–829.

Index

Accountability for ML failures, 59–60
  navigating complexity of
    accountability, 59–60
Actuarial inference, 30
Advanced encryption techniques, 63
Advanced ML algorithms, 56
Adversarial debiasing, 28–29, 41, 43
Algae biofuels, 125
Algorithmic biases, 40–41
  in education, 31
Algorithmic decision systems (ADS),
    29
  domains of, 29–35
Algorithmic systems, 29
Algorithmic transparency and bias, 54
Aliens, 178–181
  life, 178–180
  moral transition into world with,
    185–194
Antarctic programs, 216
Anthropocene, targeting of
    technologies of, 162–166
Anticipation, 178–181
Artificial General Intelligence (AGI),
    18, 85
Artificial intelligence (AI), 3, 9–10, 28,
    43–44, 52, 105–106, 113–114
  AI-based algorithms, 33
  AI-enabled robots, 72–74
  AI-powered systems, 114
  AI-powered technologies, 52
  consciousness and, 140–141
  empathy and, 141–152
  ethics, 100, 102, 149, 151
  higher education and, 144–146
  history of, 137–140
  industry, 55
  navigating coexistence of AI-
    enabled coworkers, 72–73

open questions, 151–152
rise of AI-enabled coworkers in
    manufacturing, 72
systems, 57
Artificial intelligence Fairness 360
    (AIF360), 28–29, 42–43
Artificial neural networks, 52
Artificial womb, 222–223, 225
Asimov's laws of robotics, 11, 18, 21
  application on hypothesised
    real-case contemporary
    situation, 14–18
  application on imagined future
    scenarios, 18–21
  directed at humans, 13–14
  responsibility, 12–13
Asteroid capture tech, 164
Augmented and virtual reality (AR/
    VR), 52
  challenge of privacy in, 60–62
  data privacy and security in AR/VR
    Systems, 62–64
  ethical framework for AR/VR
    technologies in industry,
    65–66
  ethical implications of AR/VR in
    industrial production, 60–66
  examples, 60–62
  psychological effects, 64–65
  psychological impacts and
    gamification in AR/VR,
    64–65
Autonomy, 65

Baseline Random Forest model, 44
Benign failure modes, 86
Biases, 28, 31–32, 40–42
Bioastronautics, 4
Bioethics, 139–140, 151–152, 224

230 Index

AI open questions, 151–152
  epharmology, 146–148
  ethics of using AI, 149–151
  first problematic intermezzo,
    140–141
  higher education and AI, 144–146
  history of AI, 137–140
  problematic intermezzo, 141–152
Biofuels, 124–127
Biomedical human enhancement, 202,
    209
Biosafety, issue of, 127–128
Biosecurity, issue of, 128–129
Biotechnology, 3–4, 173–174
"Bottom-up" approach, 98–100

Capitalism, 208–210
Case Management Inventory (CMI),
    36
Chat Generative Pre-Trained
    Transformer (ChatGPT),
    139–140, 145, 151
Civil Rights movement, 160
Classical approach, 205
Cognitive biases, 33–34
Cohen's approach, 181–182
Coherent Extrapolated Volition
    (CEV), 89
Collaborative manufacturing tasks,
    71–72
Collaborative robots (cobots), 71
Collaborative work environments,
    transition to, 72–73
Colonization, 107–108, 115
  of lifeworld, 105–106
Communicative action, 108–110
Computational framework, 28–29
Computational intelligence,
    112–113
Conference of the Parties (COP),
    120–121
Confidence, 15
Consciousness and AI, 140–141
Control, 13
Cooperative interpretation process,
    109

Correctional Offender Management
    Profiling for Alternative
    Sanctions (COMPAS),
    35–36, 39
  algorithm, 28, 43–44
  model, 40
  problems with, 42–43
  tool, 35–36
Cosmic radiation, 218
COVID-19 pandemic, 219
Criminal justice in age of AI
  algorithmic biases and debiasing,
    40–41
  domains of algorithmic decision
    systems, 29–35
  problems with COMPAS algorithm,
    42–43
  ProPublica's Study COMPAS,
    39–40
  recidivism risk assessment
    instruments, 35–39
Criticality of reliable ML, 57
Cyberphysical integration, ethical
    challenges in, 68
Cyberphysical systems, ethical
    landscape of digital twins as,
    66–68

Dartmouth Summer Research Project
    on Artificial Intelligence
    (DSRPAI), 52
Data anonymization, 64
Data gathering processes, 53
Data privacy and security in AR/VR
    systems, 62–64
  challenge of privacy in AR/VR
    landscape, 62–63
  risks of personal data misuse, 63
  securing AR/VR data, 63–64
Debiasing, 40–41
Decision-making process, 52
Deep learning models, 56, 138–139
Dehumanization concerns in
    workplace, 75
Democracy, 14
Digital technology, 105–106

Index   231

Digital twins, 52
   ethical challenges in cyberphysical
      integration, 68
   ethical framework for digital twin
      technology, 70
   ethical implications of digital twins
      in manufacturing, 66–70
   ethical landscape of digital twins as
      cyberphysical systems,
      66–68
   role of digital twins in
      manufacturing, 67–68
Digitization
   concept of digitizing human aspects,
      68–69
   of human elements and
      implications, 68–70
   potential risks and ethical concerns,
      69–70
   of social behaviour, 110
Discrimination, 32
Disparate Impact Remover, 44–45
Double-edged sword, 65
Dragon Systems from Massachusetts,
      138
Dual-process approaches, 181–182

Earth, human enhancement and
      vulnerable people on,
      205–209
Ectogenesis, 223
   potential benefits of using
      ectogenesis for space
      settlements, 222–223
Education, 31
Empathy and AI, 141–152
Energy production, 124–127
Engineering, 1
Epharmology, 3–4, 146, 148
Ethical framework
   for digital twin technology, 70
   for human–robot collaboration,
      75–76
Ethical Management of AI (EMMA),
      55
Ethical reflection, 2

Ethics (*see also* Bioethics), 114–115
   of knowledge, 129
   of using AI, 149–151
European Parliament, 11
European Union (EU), 130–131
Ex-vivo Uterine Environment platform
      (EVE platform), 222–223
Exclusion, 32
Explainable AI (XAI), 55–57
Exploitation, 202–205
Extensions, 90–93
   of prophecy, 90–91
Extraterrestrial life
   monster, 181, 185, 194–195
   moral transition into world with
      aliens, 185–194
   outer space, monsters, aliens,
      anticipation and moral
      imagination, 178–181
Extrauterine Environment for
      Neonatal Development
      (EXTEND), 222–223

Fairness, 28
Feature selection methods, 34
Federal Food, Drug, and Cosmetics
      Act (FDCA), 130–131
Federal Law on Insecticides,
      Fungicides, and
      Rodenticides (FIFRA),
      130–131
Feminist bioethics, 205–206, 211–212
Future, 1
Futures Literacy Laboratory-Novelty
      (FLL-N), 186–187

Gamification in AR/VR, 64–65
Gap, 1–2
Gender exclusion, 208–209
Gene editing, 202–203, 211–212
Generative Adversarial Network
      (GAN), 41
Generative AI, 112
Geoengineering, 172
Geosynchronous orbit (GEO),
      162–163

232 Index

Global bioethics, 205–206
Gradient Penalty, 44
Gravity fields, 220–221

Habermas, 105–106
Health service, 206
Higher education and AI,
144–146
Historical Clinical Risk
Management-20 (HCR-20),
36
Horizontal gene transfer (HGT), 124,
126, 128
Human enhancement
and type of space mission and
rationale, 203–205
and vulnerable people in space,
209–211
and vulnerable people on Earth,
205–209
Human nature, 202–203, 205
Human Resource (HR), 33
Human-centric approach, 65
Human–robot collaboration (HRC),
52
ethical framework for, 75–76
safeguarding physical and
psychological health in,
73–74
socio-economic implications of,
74–75
Humans, navigating coexistence of,
72–73
Hypothesised real-case contemporary
situation, application on,
14–18

Immune function, 221
In-process tweaks, 34
Indiana Risk Assessment System
(IRAS), 36
Industrial Revolution, 112
Industry, ethical framework for AR/
VR technologies in, 65–66
Industry 4.0 technologies, 52
Information systems, 32

Information-based 'mechanization'
process, 114
Informed consent, 224
Institutional sexism and positioning of
women's bodies as
"problem", 216–217
Instrumental convergence thesis, 88
Intelligent machines, 112–113
Intelligent manufacturing systems, 52,
55–56, 77
advent of AI and intelligent
manufacturing, 51–53
charting ethical course in intelligent
manufacturing, 76–77
dehumanization concerns in
workplace, 75
digitization of human elements and
implications, 68–70
ensuring personnel safety, 73–74
ethical considerations in machine
learning in manufacturing,
53–60
ethical dimensions of human–robot
collaboration in
manufacturing, 70–76
ethical framework for digital twin
technology, 70
ethical framework for human–robot
collaboration, 75–76
ethical implications of AR/VR in
industrial production, 60–66
ethical implications of digital twins
in manufacturing, 66–70
ethical landscape of digital twins as
cyberphysical systems,
66–68
examples, 71–72
examples in manufacturing, 66
navigating coexistence of humans
and AI-enabled coworkers,
72–73
potential job displacement and
inequality, 74–75
psychological impacts, 74
rise of AI-enabled coworkers in
manufacturing, 72

Index **233**

safeguarding physical and psychological health in HRC, 73–74
socio-economic implications of HRC, 74–75
transition to collaborative work environments, 72–73
Intensions, 90–93
of prophecy, 90–91
International Energy Agency (IEA), 125
International Genetically Engineered Machine (iGEM), 127–128
International organizations (UNESCO), 150
Interstellar craft, 164

Justice concept, 28
Justification, 169–170

Kiesling's process, 123–124
Knowledge, 108

Large language models (LLMs), 139–140
Laws of Robotics, 9–10
Learning, 148
Lebenswelt, 106–110
future, AI and ethical factor, 113–114
intelligent machines, 112–113
and technology, 110–114
unbearable complexity of, 112–113
Legacy of Socrates, 98–102
Level of Service Inventory-Revised (LSI-R), 36–37
Level of Service/Case Management Inventory (LS/CMI), 36
Liberal democracy, 173
Life, 122–123
Lifeworld, 108–109
colonization, 107–108
communicative action, 108–110
Lebenswelt, 106–110
Lebenswelt and technology, 110–114

unbearable complexity of Lebenswelt, 112–113
Liminality, 185
Local Interpretable Model-agnostic Explanations (LIME), 42
Low Earth Orbit (LEO), 162–163, 215–216

Machine learning (ML), 28, 43–44, 52
accountability for ML failures, 59–60
algorithms, 16
ethical considerations in machine learning in manufacturing, 53–60
examples, 53–54
imperative of proactive ethical engagement, 55
key ethical challenges in ML-integrated manufacturing, 54–55
models, 66, 112
reliability of ML systems, 57–59
tools, 33
transparency in ML systems, 55–57
Malignant failure modes, 86
Manufacturing
ethical implications of digital twins in, 66–70
rise of AI-enabled coworkers in, 72
role of digital twins in, 67–68
Mechanization, 112
Memex, 138
Micro-gravity, 218
Military tensions, 172
Minimal genome project, 122–123
Ministry's Code of Ethics, 10
Mission, 204–205
Monster, 178, 181, 185, 194–195
theory, 178
Monsterization, 178–180
Moral exceptionalism, 203
Moral imagination, 178–181
Moral sensitivity, 180

234 Index

Morality, 114–115
Mycoplasma mycoides, 122–123

National Aeronautics and Space
    Administration (NASA),
    216–217
National Bioeconomy Blueprint 2012
    report, The, 120–121
National space agencies, 216
Natural Language Processing (NLP),
    33
New Space, 203–204
Nonviolent Risk Assessment (NVRA),
    37
Nuclear technology, 173–174

Off-world human society, 217
Off-world pregnancy risks, 217–222
    distance from Earth, 219–220
    gravity fields, 220–221
    hostile/closed environments,
        221–222
    isolation and confinement, 219
    space radiation, 218–219
Offender Assessment System (OASys),
    37
Ohio Risk Assessment System
    (ORAS), 37
Open-source AI models, 57
Outer space, 178–181
Outer Space Treaty, The, 163

Partial merger of space skepticism with
    political activism, 168–171
Perils of synthetic biology
    artificial life from scratch, 122–123
    efficient manufacturing of
        pharmaceuticals, 123–124
    energy production, 124–127
    ethical issues of, 127–129
    history of, 121–123
    issue of biosafety, 127–128
    issue of biosecurity, 128–129
    potential benefits and ethical issues
        of, 123–129
    synthetic biology, 120–121

synthetic biology regulation,
    130–131
Personal data misuse, risks of, 63
Personnel safety, 73–74
Perverse instantiation
    aligning intensions and extensions,
        90–93
    genies once and genies in future,
        93–98
    legacy of Socrates, 98–102
    twisting wishes and instantiating,
        86–90
Perverse instantiation, 95–97
Pharmaceuticals, efficient
    manufacturing of, 123–124
Philosophical method of Socrates, 92
Physical health in HRC, 73–74
Political activism, partial merger of
    space skepticism with,
    168–171
Populism, 164
Post Conviction Risk Assessment tool
    (PCRA tool), 38
Post-processing approaches, 34
Pre-processing, 34
Pregnancy, 221
Prejudice, 32
Principles for Supervision of Synthetic
    Biology, 130
Proactive ethical engagement,
    imperative of, 55
Promises of synthetic biology
    artificial life from scratch,
        122–123
    efficient manufacturing of
        pharmaceuticals, 123–124
    energy production, 124–127
    ethical issues of, 127–129
    history of, 121–123
    issue of biosafety, 127–128
    issue of biosecurity, 128–129
    potential benefits and ethical issues
        of, 123–129
    synthetic biology, 120–121
    synthetic biology regulation,
        130–131

Index 235

ProPublica
  scrutiny of COMPAS tool, 29
  study COMPAS, 39–40
Propulsion systems, 162–163
Proto-centaur hybrids, 184
Pseudonymization, 64
Psychological health in HRC, 73–74
Psychological support mechanisms, 74
Psychopathy Checklist-Revised
      (PCL-R), 37–38

Quantitative Input Influence (QII), 45

Racial bias, 34
Radiation astronauts, 218–219
Radiation risks, 219
Random Forest
  classifier, 28–29
  model, 43
Rat neonates, 220
Rational discourse, 108
Realism, 61–62
Recidivism risk assessment
      instruments, 35–39
Recombinant DNA technology, 122
Reliability of ML systems, 57–59
  consequences of system failures, 58
  criticality of reliable ML, 57
  measures to enhance system
      reliability, 58–59
Reproduction in space, historical
      perspectives on, 216–217
Reproductive bioastronautics, 225
Reproductive bioethics framework for
      space, 223–225
Reproductive technologies, 216, 225
Responsibility, 10, 12–13
Responsibility gap concept, 13
Risk, 171–174
  communication, 224
  of personal data misuse, 63
Risk-Need-Responsivity model (RNR
      model), 36–38
Robot Ethics Charter, 11
Robotics, 165–166
Robots, 143–144

Robust access control mechanisms, 63
Rocketry, 162–163
Rossum's Universal Robots (R. U. R.),
      137–138
Royal Academy of Engineering,
      120–121

Safety, 9–10
Satellites and communications systems,
      162–163
Science and technology studies (STS),
      32
Second wave of synthetic biology, 122
Sexually transmitted infections,
      221–222
SHapley Additive exPlanations
      (SHAP), 45
Simulated liminality, unfamiliar
      futures and beneficial
      monsters in, 185–194
Situation, 109
Skeptical complex, tensions within,
      166–168
Skepticism, 160–161, 171–172
Smartphone-based gamified job design,
      64–65
Social construction of reality (SCR),
      115
Socrates, legacy of, 98–102
Socratic approach, 102
"Socratic hybrid" approach, 100
Sophisms, 91–92
Sophists, 91–93
Space, 4, 204–205
  expansion, 163
  historical perspectives on
      reproduction in, 216–217
  human enhancement and type of
      space mission and rationale,
      203–205
  human enhancement and vulnerable
      people in, 209–211
  potential benefits of using
      ectogenesis for space
      settlements, 222–223
  radiation, 218–219

236 Index

reproductive bioethics framework for, 223–225
space-based Earth monitoring technologies, 166
Space exploration, 202–203
and utilization, 180
Space skepticism, 164, 169
partial merger of space skepticism with political activism, 168–171
Space technologies, 2–3, 162–163
partial merger of space skepticism with political activism, 168–171
setting up problem, 160–162
targeting of technologies of anthropocene, 162–166
tensions within skeptical complex, 166–168
timeliness and risk, 171–174
Speech recognition models, 33
State agencies, 209–210
Static-99 Revised (Static-99(R)), 38–39
Stereotyping, 32
Super intelligence, 18–19, 87
Super-intelligent AI, 19–20, 22, 88
Super-intelligent artificial systems, 18–19
Superior Second Law, 17
Synthetic biologists, 123–124
Synthetic biology, 119–121, 129
regulation, 130–131
synthetic biology 1.0, 122
Synthetic genomics, 124–125, 129
Synthetic Minority Over-sampling Technique (SMOTE), 45
Synthia, 120, 122–123
Systems
consequences of system failures, 58
theory, 107

Technical Construction of Reality (TCR), 105–106

Technik-Lebenswelt, 105–106
Techno-constructionism concept, 105–106, 111
Technological construction of reality (TCR), 111
Technological development, 5
Technological phenomenon, 2, 115
Technological sectors, 4–5
Technology, 1–3, 105–106
Lebenswelt and, 110–114
Teleological claim, 164, 168
Temporalization, 164
Tensions within skeptical complex, 166–168
Terrestrial technology, 173–174
Tetris, 84
Theme, 111
Timeliness, 171–174
"Top-down approach", 98–100
Toxic Substances Control Act, The, 130–131
Transfer learning technique, 53
Transparency
in AI systems, 32
challenges to achieving, 56
imperative of, 55–56
in ML systems, 55–57
strategies for enhancing, 56–57
"Transparency by design" concept, 55–56
Trust, 15
Turing test of 'machine intelligence', 138
Turing's universal machine, 138

Ultra-violet radiation (UV radiation), 218–219
US National Research Council, 124–125
User-centric design, 57

Virtual reality (VR), 52
input devices, 62–63
Vulnerable group, 208, 211

Index   237

Vulnerable people in space, human enhancement and, 209–211
Vulnerable people on Earth, human enhancement and, 205–209

War, 160
Wasserstein Generative Adversarial Networks (WGANs), 44
Women, 206–207

Women's bodies as "problem", institutional sexism and positioning of, 216–217
Workplace, dehumanization concerns in, 75
Workshop participants, 188

Xenobots, 152

Zeroth Law, 9–11, 15–16, 18

Printed and bound by CPI Group (UK) Ltd, Croydon, CR0 4YY
05/02/2025
14638643-0002